Homeopathy in Perspective

Anthony Campbell was consultant physician at The Royal London Homeopathic Hospital (now The Royal London Hospital for Integrated Medicine) for over twenty years. He retired in 1998. He has published books and articles on homeopathy, medical acupuncture, and complementary medicine as well as three books on back pain for patients. Since retiring he continues to teach medical acupuncture to health professionals. He has many interests outside medicine and has published a novel as well as several books on non-medical topics.

Also by Anthony Campbell

All You Need to Know About Acupuncture
Medical Acupuncture: A Practical Guide
So You Want to Try Acupuncture?
The Assassins of Alamut
Totality Beliefs and the Religious Imagination
Religion, Language, Narrative and the Search for Meaning
Making Word.doc Files on Linux

HOMEOPATHY IN PERSPECTIVE

A Critical Appraisal

Anthony Campbell

Typeset on ArchLinux using L$_Y$X

Contents

Preface

This is is a book about homeopathy: what it is, how it developed, where it stands today. It is written for people with questioning minds; anyone who has adopted a fixed opinion in advance about homeopathy, either for or against, may receive the odd shock.

No prior knowledge of homeopathy is assumed, but this doesn't mean that the book is intended only for beginners. Even if you have read a good deal about homeopathy previously you may, I believe, find that you view it in a different light after you have finished. What I try to do here is to look at homeopathy as squarely as possible and to provide the facts as I see them. What you then make of them is up to you. I think this is worth doing because the material I present is not well known even to many homeopaths, yet it's essential for anyone who wants to make an informed judgement about homeopathy. These facts are not, to my knowledge, easily available anywhere else; it certainly took me a long time to learn them.

The book is in two parts. Part I explains the origins of homeopathy and how it came to be where it is today. It provides explanations for some of the more outlandish ideas in homeopathy and sets them in their historical context, without which it is not really possible to form an adequate idea of the subject. Part II is concerned with homeopathy now and considers the evidence for its effectiveness. I do this by looking at the research evidence and also by describing my own experience, over 24 years, of actually using the treatment.

If you want to get a proper understanding of homeopathy you should read both parts. If you are mainly interested in evaluating it from a scientific point of view you could start with Part II, although you may need to refer back to Part I if some of the ideas and terminology are unfamiliar to you. These are listed in the Index.

I thought it best not to burden the text with footnotes in the style of a scholarly monograph, but the appendices provide references and reading lists for those who want to take their study of the subject further.

A Note on the Revised Edition

There are no major changes in the text for this edition. My view of homeopathy has not altered substantially since I last wrote about it six years ago, although my scepticism about many of the claims made for it has if anything increased and this has prompted a little rephrasing here and there. I have also updated some of the information to reflect changes that have occurred, such as the closure of some homeopathic hospitals in Britain and the renaming of The Royal London Homeopathic Hospital as The Royal London Hospital for Integrated Medicine. There is one new illustration and the index has been improved.

Part I

ORIGINS AND DEVELOPMENT

Chapter 1

Introduction

Homeopathy is under attack today, at least in Britain. There is nothing new about that, of course; throughout its 200-year history it has been repeatedly criticised as inherently absurd and useless, yet it has survived. The National Health Service currently funds homeopathic treatment by GPs. Until recently there were five NHS homeopathic hospitals in Britain; some have now closed but the Glasgow Homeopathic Hospital still exists while The Royal London Homeopathic Hospital has been renamed The Royal London Hospital for Integrated Medicine and continues to offer homeopathy along with other complementary treatments.

Critics object that in a time of limited resources and ever more expensive conventional treatments it is unjustifiable to fund this support for what they regard as a discredited form of medicine. True, there is still a lot of demand for it from patients, but funding health authorities are asking for evidence of efficacy and this is not easy to provide. The media, which until recently were pretty uniformly sympathetic to homeopathy, are beginning to be less so or even downright hostile. All this is likely to seem puzzling to people who know little about the subject. In this book I try to present the facts as opposed to the rhetoric.

A brief overview of the territory

Homeopathy is a system of medicine that was introduced in the early nineteenth century by a German physician, Samuel

1

Christian Hahnemann (1755–1843). It was based on the idea
of "like curing like". Later, Hahnemann also introduced the
use of very small doses, which he came to believe were actually
more effective than larger ones; this is the feature that has most
strongly captured the popular attention.

Homeopathy spread widely in Europe in the 19th century
and was brought to England and also to the USA, where it
became very successful after the Civil War. In the USA it took
on a different character, when it was coloured by ideas derived
from the Swedish mystic Emanuel Swedenborg. Towards the
end of the nineteenth century an American homeopath and
Swedenborgian, James Tyler Kent, was very influential and his
ideas were brought to Britain, where they became the dominant
homeopathic orthodoxy after the First World War.

Kentian homeopathy subsequently was exported all over
the world and is still widespread today. It is characterised by
hostility to orthodox medicine, the use of very dilute medicines
("high potencies"), and emphasis on the psychological and
"spiritual" characteristics of patients. Many of the more extreme features of modern homeopathy can be ascribed to Kent.
Other forms of homeopathy do however exist: for example,
"complex homeopathy", much used in Germany, where it tends
to shade off into herbal medicine (phytotherapy).

There is a sense in which homeopathy could be thought
of as a kind of medical coelacanth, a survival from an earlier
age. Still, it has had to move with the times and this means
accepting the need to test the treatment objectively by means
of controlled trials. These have been done to some extent and
have given mixed results. In spite of continuing uncertainty
about its efficacy, homeopathy answers a need felt by many
people and for this reason alone it is likely to continue to be
used in the twenty-first century.

A typical homeopathic consultation

A homeopathic consultation generally differs quite a lot from what
goes on in a conventional doctor's surgery. Suppose you go to see
your GP with what you suspect is arthritis. She will ask you a

number of questions with the aim of reaching a diagnosis. She will want to know how long you have had the symptoms, how they began, which joints are affected, and so on. She will examine your joints and probably your other systems as well — heart, lungs, skin, for example. She will arrange for some blood tests. At the end of this process she will reach a diagnosis, perhaps rheumatoid arthritis. She will then prescribe some treatment or, if necessary, refer you to a rheumatology specialist for a further opinion.

What has been happening throughout all this is a process of elimination. The doctor is excluding other causes for your symptoms until she is left with rheumatoid arthritis as the only reasonable conclusion. This is quite similar to what a detective does in solving a crime — putting together facts, making hypotheses, testing them to see if they are valid. The science or art of making a diagnosis in this way is one of the main things taught at medical school.

A homeopathic consultation, in contrast, is not concerned with diagnosis in this sense, although if the homeopath is medically qualified she will have made a conventional diagnosis in addition. But the essence of homeopathy is said to be *individualisation*. Instead of seeking to place you in a class (the rheumatoid arthritis class in this case) the homeopath is interested in those features that distinguish *you* from other patients with broadly similar symptoms. This is in many ways the opposite of what happens in a conventional consultation.

For example, she will ask about things that make your pain better or worse — heat, cold, movement, and about your emotional reactions to your symptoms — whether you get irritable, weepy, fearful, or worried. She will also ask a range of general questions about your appetite, sleep, reactions to weather, food likes and dislikes, and other things that have nothing directly to do with your complaint. Here she is using the homeopathic concept of "constitution"; many though not all homeopaths believe that patients can be classified in a number of groups according to their personalities and that this may give an important clue to the kind of medicine, or remedy, they require.

Obtaining all this information can take a long time — up to an hour or even longer. Armed with it the homeopath will now try to decide on the right medicine for you. In the past this often entailed referring to a large book, called a repertory, in which various symptoms are listed according to the remedies that correspond to them. As there are often hundreds of these for a particular symptom a lot of complicated sifting and cross-referencing is called for, and the task has been computerised to make it easier and quicker. Several programs are available for this purpose and many homeopaths now make use of them.

Having decided on what she hopes will be the right medicine ("remedy") the homeopath will prescribe it in one of several forms. It may be a liquid, tasting of alcohol; a sugary powder, to be dissolved on the tongue; or little globules of sugar. You may be asked to take just a single dose of the remedy, though this may be split in three to be taken over twenty-four hours, or you may be given a number of tablets to take once or twice daily for several weeks; other ways of taking the medicines are also possible. In any case, you will probably be asked to return in a few weeks to report progress.

You will probably be warned that your symptoms may become worse at first; this is the so-called homeopathic aggravation, which is generally thought to be a good sign — the right medicine has been chosen. You may be advised to avoid coffee and tea and told to keep the medicines away from strong light or scent. If you ask how homeopathy works you will be told that it stimulates the body to heal itself and that it is "getting at the root" of the problem.

When you return for your second consultation you will be asked how you got on. The homeopath will be particularly interested in any changes in your mood, appetite, and particularly your general sense of well-being. Depending on what has happened she will decide whether to repeat the remedy, change to a different one, or do nothing. The last possibility may well surprise you but the homeopath will explain that it is a principle of this form of treatment not to repeat the treatment until its effects begin to wear off. Many homeopaths believe that remedies can continue acting for weeks or even months. Homeopaths warn their patients that treatment of chronic disease is likely to be lengthy and quick results are often not to be expected.

The description I've just given applies to chronic disease, such as rheumatoid arthritis. Homeopathy originally developed as a treatment for acute disease, however, and in this case the consultation need not be so long or so detailed. An acute sore throat, for example, is often treated with one of a quite small range of remedies and the homeopath may be satisfied to prescribe on just a few presenting symptoms in these circumstances.

If you are encountering homeopathy for the first time you will probably feel that you have received a gratifying amount of individual attention but you are also likely to find it pretty mystifying. That was certainly my experience.

My background in homeopathy

I first became involved with with homeopathy in 1974, when I began working and studying at The Royal London Homeopathic Hospital (RLHH). At the time I had just completed a period of intensive study to obtain membership of the Royal College of Physicians and was looking about for a new direction professionally. Even so, I knew nothing about homeopathy and would not have thought of taking it up, were it not for a friend whose opinion I respected and who told me she had found it to be impressively effective.

In part, homeopathy appealed to me because it was different. Conventional medicine in the early 1970s seemed not to be going forward very fast. Little new was happening and medical advances, such as they were, consisted mainly in ringing the changes on the existing antibiotics and antidepressants. If homeopathy really worked, as my friend assured me from her own experience it did, it seemed to offer a new and potentially exciting alternative.

I therefore registered to take an introductory course at the hospital. Here I quickly found myself floundering in unknown waters. The names of the medicines were nearly all unfamiliar and I thought I must have misheard when a lecturer mentioned casually that she had used a more dilute preparation in order to obtain a bigger effect. These things were disturbing enough, but what really worried me was the feeling that taking up homeopathy might be a little like undergoing a religious conversion. At least some of the speakers seemed to feel themselves to be in possession of a sacred Truth that had to be preserved at all costs; one man reacted quite angrily when someone in the audience expressed a degree of scepticism about something he had said.

Fortunately, not all the homeopaths I met were so closed-minded. And there were definitely career prospects available to someone like me, for the homeopathic community was in a state of crisis at this time. In 1972 sixteen doctors and senior staff from the hospital had been killed in an air crash when taking off from Heathrow on their way to a conference in Brussels. Although the victims had not included any of the consultants, at least two of these were close to retirement and there was deep concern about the need to find replacements. As I had the higher medical qualification needed to be appointed as a National Health Service consultant I was a potential recruit to the homeopathic banner.

The RLHH at that time was a small general hospital, with a very busy outpatients service and well over a hundred inpatients, though more of these were surgical than medical. There were several surgeons working at the hospital and they were in no sense homeopathic, although surgical patients sometimes asked for a consultation with one of the homeopathic physicians.

On the medical side the hospital had a slightly old-fashioned air, which was accentuated by the fact that several of the physicians were elderly. However, there were also some younger staff members and all of them, old and young, gave the impression of being good doctors as well as good homeopaths. In fact, they were doctors first and homeopaths second, which was definitely reassuring, though this is not to say that homeopathy was unimportant to them. It seemed clear to me that, as they claimed, practising this form of treatment entailed listening to patients in a way that the patients themselves appreciated and that had to some extent died out in mainstream medicine.

In view of the succession problem the hospital had recently set up a training scheme for future consultants which combined clinical work and study. I was appointed to this scheme along with another doctor a few years younger than I. In 1977 several consultant vacancies arose owing to retirement and resignation, and a colleague and I took over as consultants. I remained at the hospital until my retirement twenty-one years later.

My opinion of homeopathy naturally evolved and changed over the years I was at the hospital. After an initial period of enthusiasm it became gradually more questioning and ultimately more critical.

Early successes were not repeated and I increasingly began to suspect that the responses I did occasionally see were due to the placebo effect or other things rather than to the medicines. I attended talks or read articles in which people described their successes with various homeopathic remedies but they never seemed to work for me.

When I first encountered homeopathy I was taught what is often called "classical homeopathy". For reasons I shall explain in the course of the book, I think this is a misleading term, but at the beginning I didn't realise that there were other ways of practising homeopathy. Nevertheless I was puzzled by many of the ideas I encountered, some of which seemed quite bizarre. Where did these come from? This never seemed to be explained. Many homeopaths assumed that they all originated in the nineteenth century with the founder of homeopathy, Samuel Hahnemann, although they also attributed great importance to a later American writer called James Tyler Kent. But most were busy working physicians with little interest in delving into the origins of their discipline; most were content to rely on hearsay and secondary or even tertiary sources.

One of my tutors, however, was an erudite and widely read doctor called Ralph Twentyman. He told me that Kent had been much influenced by the eighteenth-century mystic Emanuel Swedenborg, and this remark made me aware of the need to go back into the origins of homeopathy in order to try to understand it better. In the 1970s and 1980s I began to read the writings of British nineteenth-century homeopathic physicians such as Robert Dudgeon and Richard Hughes. These authors revealed to me a picture of homeopathy that was significantly different in many ways from that with which I was familiar. In their books I found explanations for many aspects of homeopathy that seemed eccentric or outlandish, and I began to publish my discoveries in papers and, later, in book form.

The book, which I called *The Two Faces of Homeopathy*, came out in 1984. It sold reasonably well and stirred up quite a bit of controversy, but it is now out of print. Since then new facts have come to light and my own ideas about homeopathy have developed further, so this book is a very substantial update of the earlier work. But one thing has remained the same. In both books my approach is mainly historical. This requires a few words of explanation. Why this interest in history? Isn't present-day homeopathy what really matters?

The importance of history

One of the many ways in which homeopathy differs from conventional medicine is in the importance of history for understanding it. In conventional medicine the history of the subject is fairly unimportant; it has a cultural value but medicine changes so fast that the ideas of the past have little relevance to the present. For homeopathy it is otherwise. *I'd go so far as to say that you can't understand homeopathy in any depth unless you have a fair idea of its history.* Yet this fact is not always recognised by newcomers to the subject, who in consequence find themselves more confused than they need be.

Like many other forms of unorthodox medicine, homeopathy is largely static, fixed in the past; most of the ideas still current today have altered little from when they were first formulated in the nineteenth century. The main textbooks still in use today were written at that time. For homeopaths this is not usually thought of as a disadvantage — rather the opposite. Believers in unconventional medical systems often regard their perceived antiquity as a merit. In the case of homeopathy there are claims that it has ancient roots, even as far back as Hippocrates.

Important though the history of the subject is, however, homeopathy is still a living form of treatment and the book would be incomplete if it were merely a historical study. In Part II, therefore, I look at homeopathy today and consider the all-important question: does it work? For reasons I shall explain, this is actually quite difficult to answer, partly because of the lack of much good-quality research but also because it isn't easy to define homeopathy. But I shall do my best to offer a balanced assessment, based partly on research but also on first-hand experience.

The practical issue: should I try homeopathy?

This brings me to a difficult but important question, which was raised by a perceptive reviewer of an earlier draft of this book. What if you, the reader, want to know whether you should try homeopathy yourself?

In her critical and witty book *Sleeping With Extraterrestrials* the sceptical writer Wendy Kaminer begins by admitting that she consults a homeopath. She is embarrassed by this, but nevertheless she finds that, for whatever reasons, homeopathy has helped her.

> When I go to my homeopath maybe I'm following one of the precepts of the recovery movement that I've always derided: I'm thinking with my heart and not my head. Or maybe I'm acting rationally after all. Believing in homeopathy may be irrational, but not using homeopathy if it works would be even more irrational. I care only if medicine works, not why. (I have the vaguest understanding of antibiotics.)

> So I don't listen to scientists eager to tell me why homeopathic remedies can't possibly work, because they violate the laws of chemistry. Assuming that the scientists are right, and the remedies I've taken are mere placebos, why would I want to start doubting — and diminishing — their effectiveness? Why not be susceptible to placebos?

Kaminer surely has a valid point here: it may be rational for *an individual* to use homeopathy, even if the benefit is due to the placebo effect. But, as she goes on to point out, it would be irrational for *anyone else* to take her belief in the efficacy of homeopathy at face value; she might be mistaken, deluded, or even dishonest:

> If you're intrigued by my report, you should ask me to substantiate it, with some objective evidence. You should try to duplicate my experience.

It is, I find, remarkably difficult to find much "objective evidence" about homeopathy; frequently what we get instead is emphatic assertions with few supporting facts. In this book I try to present the available evidence as fairly as I can; it is for you, the reader, to make of it what you will.

Chapter 2

Hahnemann's Discovery

There is a prevailing view of the origins of homeopathy that is partly based on legend: it might be called "homeomythology". The legend goes something like this.

> *Samuel Hahnemann, the founder of homeopathy, was a medical genius whose thought was far in advance of his own time and even ours. He started out as a conventional doctor but became disillusioned with the orthodox approach, which he called allopathy, and cast about for something better. As the result of an experiment on himself using quinine he was led to formulate the homeopathic "law of similars". He also discovered the principle of using small doses.*
>
> *In the light of these two ideas he carried out a vast number of experiments on himself and others (using tiny doses, naturally) — the so-called "provings". These are the basis of his system. Modern homeopathy still relies on the provings carried out by Hahnemann and his successors down to our own day. It is often also believed that Hahnemann introduced the idea of prescribing on the basis of the patient's character or "constitution".*

Like most legends, this one is based on fact but it also contains fantasy, and it incorporates ideas that were not part of Hahnemann's own doctrine but were introduced from other medical or even mystical belief systems of the time. However, although the commonly accepted idea of how homeopathy originated contains much that is legendary, in one respect at least it is accurate: its originator, Samuel Hahnemann, is at centre stage. If, as has been said, Western philosophy is a series of footnotes to Plato it is even more true that homeopathy is a series of footnotes to Hahnemann. My first task, therefore, must be to present an outline of Hahnemann's life and thought.

Samuel Christian Hahnemann was by any standards a glorious eccentric, and his restless life story is mirrored in the turbulent history of the medical heresy that he fathered. In order to understand him and his views we must set him in his historical context, for his life and career span a critical period in the development of European medical and scientific thought, in which ways of looking at the world and at human beings that still owed much to classical and mediaeval ideas were giving way to those with which we are familiar today. This is reflected in Hahnemann, who at times seems almost modern and at others appears to be living in a conceptual universe so remote from our own as to be scarcely comprehensible.

Life

Hahnemann was born at Meissen, in south-east Germany, on 10th April 1755, at approximately midnight. So, at least, Hahnemann himself always maintained; but the entry in the church register at Meissen records the birth as having occurred on the morning of 11th April, and this later date was adopted by some homeopaths and gave rise to disagreement about the right day to celebrate the Master's birthday. It is curiously appropriate that the inventor of homeopathy should have arrived in this world already equipped with a future occasion for controversy.

Hahnemann's father was a craftsman who worked in the famous Meissen porcelain trade. He was not very well off, so as a boy Samuel was briefly put to work for a Leipzig grocer, and it was with some difficulty that he persuaded his father to allow him to become a medical student. But in 1775 he entered the University of Leipzig, where he quickly became self-supporting by means of teaching and translating. Growing dissatisfied with the standard of medical education at Leipzig, however, and already showing signs of a characteristic restlessness, he departed in 1776 for Vienna, but before completing his studies there he left to take up a post as librarian and family physician to the Governor of Transylvania, Baron von Brukenthal, at Hermannstadt.

Figure 2.1: Samuel Hahnemann

It was at this time that he became a Freemason. It has been claimed that the library at Hermannstadt held esoteric alchemical works, including those of Paracelsus, and that it was dipping into these that planted the seed of homeopathy in Hahnemann's mind. This is certainly possible, but no evidence to support the speculation exists.

In 1779 Hahnemann left his employment with von Brukenthal to complete his medical education at the University of Erlangen, where he was finally awarded his doctorate in medicine in August 1779. We don't know what Hahnemann did in the year after qualifying, but in 1780 he established his first medical practice in the small mining town of Hettstedt, where he recorded his disillusionment with the medical treatments of his day, especially bloodletting.

Soon afterwards he moved to Dassau, where he began to take an interest in chemistry. This was an momentous period for chemists. In Hahnemann's lifetime the phlogiston theory of combustion was disproved, a number of gases were identified, the compositions of air and water were discovered, and the atomic theory was placed on a surer footing. Hahnemann felt the excitement of this atmosphere of discovery and carried out some chemical research of his own, though he doesn't seem to have heard about the atomic theory.

The sense of intellectual excitement was not confined to chemistry. At this time German writers and philosophers were developing the ideas of "Naturphilosophie" — a semi-mystical view of science and the world that underlies, for example, much of Goethe's thought. It is almost certain that the young Hahnemann would have encountered these ideas at university, and though he does not refer to them explicitly in his writings their influence can be detected in his religious and metaphysical outlook. Naturphilosophie, as expounded by its leading philosopher, Schelling, is based on a sense of the Divine as underlying the manifest universe and as giving form to it, but the nature of truth is to be apprehended by thought and intuition rather than through revelation.

Naturphilosophie, therefore, is deistic rather than theistic, pagan rather than Christian. Hahnemann, likewise, though a deeply religious man who believed himself to be God's chosen instrument for the healing of mankind, was hardly a Christian: nowhere in his writings does he refer to Christ or Christianity. His religion is essentially a matter of faith in, and devotion to, the Father. This religious attitude found itself at home in Freemasonry; Hahnemann preserved his interest in the Craft all his life, though he was not always an active member.

In 1782 he married Johanna Kuchler, an apothecary's daughter. A year later their first child, a daughter, was born — the first of a large family. Still Hahnemann did not settle down but continued to move about. In 1785 he was in Dresden, where he worked as locum tenens for the Medical Officer of Health. On the death of the incumbent Hahnemann applied for the substantive appointment, but he was unsuccessful and set off once more on his travels.

He seems then largely to have abandoned medical practice for a time and to have concentrated his energies on translating, by which he supported his family and himself for a number of years. He also continued his chemical research; he published a test for the fraudulent adulteration of wine with lead which was officially adopted in Prussia, and he described a method for detecting arsenic in forensic material. It is said that while in Dresden he met the famous French chemist Lavoisier, later to be guillotined during the Revolution.

In 1789 Hahnemann and his family moved to Leipzig. This was Hahnemann's third sojourn in that city. He did not practise medicine there but continued to write, translate. and study. His family now consisted of six persons, and he found himself hard pressed financially.

There is a touching story that gives a vivid picture of the tribulations undergone by the Hahnemann family. At one time money was so short that Hahnemann used to weigh out a portion of bread daily for each member of his family. When one of his daughters fell ill she was unable to eat her ration, and so stored it away in a box until she should recover. But she began to feel worse rather than better, and fearing she would die she called her favourite sister and handed over to her the store of dried-up bread as a legacy so that it should not be wasted.

To have grown up in the Hahnemann household seems to have been something of an ordeal in various ways and it left its mark on those who underwent the experience. The family was dogged by tragedy. Two daughters were probably murdered and three were divorced, while the elder son Friedrich seems to have been half-mad. He deserted his wife and child; his ultimate fate is unknown, but there is a curious story of a wild-looking man called Hahnemann who appeared in America during a cholera epidemic, cured a large number of people, and then vanished into the far West, never to be seen again; this was probably Friedrich.

Hahnemann's other son died as an infant in 1799, when Hahnemann was forced to leave Koenigslutter owing to the hostility of the pharmacists of that town (a harbinger of things to come). On the way to Hamburg the carriage in which the family was travelling was overturned; Hahnemann's son received fatal injuries and one of his daughters broke a leg, so that the party had to interrupt the journey for over six weeks.

The role of Frau Hahnemann amid all these vicissitudes is uncertain. No doubt she had a difficult life, but there are suggestions that she was something of a Xanthippe to her philosophical husband. In the circumstances it is perhaps hard to blame her.

Between 1789 and 1805 the Hahnemann family lived in literally dozens of places in eastern Germany. Hahnemann was unable to settle anywhere, but was driven on by his restless spirit and the need to make a living. All this travelling was a more difficult, indeed hazardous, affair then than it would be today. The roads were bad and often unsafe, and moreover the period was one of continual civil unrest. Hahnemann's youth was marred by the Seven Years' War between Prussia and Austria, while later, at Leipzig, he was to find himself caught up in the Napoleonic Wars.

Although Hahnemann was not practising medicine at this time he still had strong views on the subject, which he repeatedly expressed forcibly in print. The prevailing medical theories of his day were based on crude mechanical and hydraulic analogies as explanations of physiological processes. Thus diseases were classified in terms of tonicity or relaxation (our use of the word "tonic" derives from this theory) or were ascribed to supposed intestinal inflammation. There is no need to discuss these long-discredited theories in detail but it is important to notice their practical implications for medical treatment.

The main resources of orthodox physicians in Hahnemann's day were large doses of drugs, habitually given in complicated mixtures, and blood-letting, often carried to horrifying lengths — indeed, to the point of almost complete exsanguination, so that the final drops had to be squeezed from the unfortunate patients.

Hahnemann rejected both the theories and the practices of orthodox medicine. It was, he held, inherently impossible to know the inner nature of disease processes and it was therefore fruitless to speculate about them or to base treatment on theories. As for complex drug mixtures and blood-letting, both were dangerous and unjustifiable.

Hahnemann had not yet thought of homeopathy but he was a firm advocate of environmental measures to promote health — fresh air, good food, and exercise. In these opinions he was certainly in advance of his time, and the same is true of his enlightened ideas about the right way to treat the mentally ill.

In Hahnemann's day the practice was to treat "lunatics" with great harshness; they were given purges and emetics and were tied up, starved, and flogged if they complained, soiled themselves, or became violent. Hahnemann strongly attacked this crude form of behaviour therapy and instead advocated kindness and patience. In 1792 he had an opportunity to put his ideas into practice, for he was invited by the Duke Ernst von Sachsen-Gotha to come to Georgenthal to manage an asylum for the insane. The Duke magnanimously placed part of his hunting-castle at Hahnemann's disposal for the purpose.

Unfortunately only one patient was ever admitted. This was a Hanoverian government minister named Klockenbring. Hahnemann left his patient free and gradually built up a rapport with him; he also gave him medication though we do not know much about this. Under this treatment Klockenbring recovered and was discharged, though he relapsed and died two years later. After this no new patients were forthcoming and the experiment came to an end. There is a suggestion that Hahnemann, always a difficult man to get on with, had fallen out with his patron the Duke. He had certainly gained a reputation as an eccentric: the Sheriff of Georgenthal, when asked how many patients Dr Hahnemann had in his institution, replied drily: "One — himself."

Hahnemann therefore recommenced his wanderings. His family now consisted of ten persons and financial pressures were greater than ever. He tried to support himself by admitting mental patients to his home, but this was not a success and he had to fall back on his old trade of translating.

He also made two unwise attempts to remedy his fortunes by other means. In 1800 he published an announcement of his discovery of a cure (belladonna) for scarlet fever, which he promised to reveal to anyone who paid a gold piece for his book on scarlet fever; and in 1801 he mistakenly believed that he had discovered a new chemical compound of possible medicinal value, details of which could be obtained on payment of a fee. It proved to be common borax. These unprofessional announcements earned Hahnemann a good deal of derision and opprobrium.

Up to this time Hahnemann had done scarcely enough to earn even a footnote in medical history. By now he was into middle age and it might have seemed likely that he would never again have anything to do with practical medicine, but in fact his real medical career was about to begin.

The discovery of homeopathy

The germ of homeopathy had been planted in Hahnemann's mind by an experiment he carried out in 1790. It was suggested to him by translating the *Materia Medica* of the Scottish physician Cullen. Among the herbs described by Cullen was the Peruvian bark cinchona (quinine), already in use as a treatment for malaria.

Cullen followed orthodox opinion in attributing its effectiveness to its "tonic effect on the stomach". Hahnemann (who was never content to remain a mere translator but frequently added his own opinions in notes) attacked this idea, on the reasonable grounds that the taking of much more astringent substances than cinchona did not cure fever; hence the therapeutic effects of cinchona must be produced in some other way. Not content to leave the matter at the level of theory, Hahnemann proceeded to experiment.

> I took for several days, as an experiment, four drams of good china (cinchona) daily. My feet and finger tips, etc., at first became cold; I became languid and drowsy; my pulse became hard and quick; an intolerable anxiety and trembling (but without a rigor); trembling in all the limbs; then pulsation in the head, redness in the cheeks, thirst; briefly, all those symptoms which to me are typical of intermittent fever, such as the stupefaction of the senses, a kind of rigidity of all joints, but above all the numb, disagreeable sensation which seems to have its seat in the periosteum over all the bones of the body — all made their appearance. This paroxysm lasted for two or three hours every time, and recurred when I repeated the dose and not otherwise. I discontinued the medicine and I was once more in good health.

Critics have objected that quinine does not in fact produce the symptoms of malaria, but this seems rather beside the point. What matters is that Hahnemann believed that it had done so in his case and that this suggested the idea of homeopathy to him. (The clinical thermometer had not been invented in his day, so the diagnosis of "intermittent fever" was necessarily based entirely on the symptoms.)

Nevertheless, many years were to elapse before the germ of homeopathy grew into a full therapeutic system. Not until 1796 did Hahnemann publish anything on the subject, and even then the essay he wrote was theoretical rather than practical and it seems that he had not yet had much opportunity to try his idea out on patients.

In 1805, after several more moves, Hahnemann settled for a time in Torgau, on the Elbe, where he remained for an unwontedly long time — nearly seven years. We know little about his life at this time, but it seems he was practising medicine according to his new system. His finances now improved and he was at last able to give up translating and concentrate on his own writing. Numerous articles by him appeared, the most important of which was an essay, *The Medicine of Experience*, which came out in 1806 and was the forerunner of his definitive theoretical work, *The Organon*.

The Medicine of Experience was published, like many of Hahnemann's writings, in *The Journal of Practical Medicine*, edited by Hufeland — an eminent physician who, though he never became a homeopath, was sympathetic to Hahnemann's ideas. Although Hahnemann did not use the word homeopathy in print until the following year, we find set forth in this essay the main features of his method, which may be summarised as follows.

- Medicines are to be chosen on the basis of the patient's symptoms, without reference to the supposed disease process underlying them. For Hahnemann the symptoms *are* the disease, and once they have gone the disease is cured.

- The effects of drugs can be known only by means of experiments on *healthy* people. It is no use relying on what is found in patients because the symptoms of the disease will be difficult to distinguish from those of the drug.

- Medicines must be chosen for the similarity of their effects to the symptoms of the patient. This "similia principle" is of course the kernel of the homeopathic method.

- Medicines are to be given in single doses instead of complex mixtures.

- Medicines are to be given in small doses to prevent "aggravations". (Hahnemann believed that a correctly chosen medicine would always produce some slight worsening of the patient's condition, no matter how transient; this could be reduced to a minimum by judicious reduction of the size of the dose.)

- Medicines are to be repeated only when recurrence of the patient's symptoms indicates the need.

These principles constituted homeopathy as it stood when first formulated by its originator. As a system it was very different from the orthodox medicine of the day but from a modern point of view it could fairly claim to be more scientific and certainly a lot safer. At any rate, it quickly brought success to Hahnemann, who was henceforth not to find himself again penurious. What is remarkable is that he had taken some fifty years to arrive at his system, and he was to go on adding to it almost up to his death in his eighty-ninth year. He was indeed a late developer.

As well as *The Medicine of Experience*, Hahnemann published while in Torgau a book, in Latin, on pharmacology. In it he described 27 drugs, giving the symptoms they produced in the healthy body. It seems he had already tested the drugs on himself and on his long-suffering family and the book is therefore the first published record of "provings" (the testing of drugs on healthy people).

Unfortunately he gave no details about the doses he used or the manner of administration, a reticence that was to characterise all his later writings and to detract from their value. Among the drugs described by Hahnemann were aconite (monkshood), arnica (leopard's bane), belladonna (deadly nightshade), chamomilla (camomile), nux vomica (poison nut), and pulsatilla (windflower), all of which are still widely used in homeopathy today.

In 1810 Hahnemann published the first edition of his major theoretical work, *The Organon of Rational Healing* (later retitled *The Organon of the Healing Art*, and today often referred to simply as *The Organon*). Further editions of this continued to appear at intervals throughout his long life, while the sixth and last did not come to light until 1920.

The Organon is the Bible of homeopathy and anyone who wants to study the subject seriously must read it with close attention — a somewhat daunting task. It is arranged in numbered paragraphs, to which are often appended voluminous footnotes. The style is difficult — long involved sentences that the most authoritative English version, that of R.E. Dudgeon, does not render wholly pellucid.

In the course of his life Hahnemann was to have second and third thoughts about many of the ideas in *The Organon*; these he incorporated in the text of each successive edition, though without always cancelling what he had written previously, so that self-contradictions occur. Coming to terms with Hahnemann's thought therefore involves the reader in some fairly detailed textual criticism, and it is not surprising, if regrettable, that many later homeopaths have shirked the task and consequently have had an over-simplified view of what the Master actually taught.

The Organon initially excited rather little interest, either hostile or friendly. Perhaps this was because of distractions from public events, for the Napoleonic Wars were now raging. Napoleon himself entrenched outside Dresden in the winter of 1810–11 and constructed fortifications at other towns, including Torgau, further down the Elbe. Feeling understandably unsettled by these preparations for war, in 1811 Hahnemann decided to move to Leipzig; an unwise choice as it turned out, for Leipzig was to become the site of one of the most decisive battles of the war.

This was the fourth time that Hahnemann had stayed in Leipzig; the first time had been as a grocer's boy, the second as a medical student, and the third as a struggling physician. None of these visits was a happy precedent, but on this occasion — at least to begin with — things went better for him.

Fame at last

Hahnemann's first venture was to try to set up an Institute for the Postgraduate Study of Homeopathy, but no physicians enrolled for the course and he therefore applied to be allowed to deliver lectures at the university. Candidates for this honour were expected to present a dissertation and to defend their thesis in the mediaeval fashion against a "respondent". With unwonted tact Hahnemann avoided the contentious subject of homeopathy and instead presented a learned paper designed to prove that the white hellebore of the ancients was the same as the modern Veratrum album. The respondent was his son Friedrich. The subject proved acceptable, the occasion went off well, and Hahnemann was free to begin his lectures.

In the same month Napoleon began his calamitous retreat from Moscow. By August 1813 he was back in Saxony with a new army; he defeated the allies at Dresden and then moved north-west to Leipzig, where he encamped outside the city accompanied by his unreliable ally the King of Saxony.

On the 18th October Napoleon fought a major battle against the Allies, who were commanded by Prince Karl Schwarzenberg. Next day Napoleon's Saxon allies turned against him; he was defeated and had to leave Germany, never to return. Leipzig celebrated the defeat of the French but the city was full of wounded men. Hahnemann took part in treating the casualties and the victims of the epidemic that broke out in the city.

Gradually life in Leipzig returned to normal and Hahnemann was able to resume his lectures. At first these were packed, large numbers of students turning out for what they expected would be a rag occasion. Hahnemann himself took matters with extreme seriousness but even his closest friends and disciples felt that the solemnity of the setting left something to be desired.

Hahnemann, his few remaining white hairs carefully curled and powdered, and wearing formal clothes that belonged to a bygone era, would sit down ceremoniously, take out his watch and lay it before him on the table, and after clearing his throat read a passage from *The Organon*. He would then dilate upon the ideas it contained, becoming more and more excited and flushed, until at last he broke out in a "raging hurricane" of abuse against orthodox medicine and orthodox practitioners. This, of course, was what his audience was waiting for.

Once the entertainment value of the lectures had been exhausted, however, attendance dwindled and soon Hahnemann was reduced to lecturing to a few devoted disciples. But his lack of success was not due solely either to his subject matter or to his eccentricities of dress and delivery; he was the target of serious opposition from the Professor of Medicine, and even those students who would have liked to come over to the new system of therapy found it unwise to do so.

Yet if Hahnemann failed to make his mark as a lecturer his sojourn in Leipzig was immensely fruitful in another way, for it was at this time that he carried out his main series of "provings" with the help of his few disciples.

The little band of enthusiasts was worked hard by the Master. Not only did they have to try out the various drugs on themselves and record the results with extreme conscientiousness; sometimes they had to collect the substances, especially the herbal ones, themselves, learning to recognise them in the field and to prepare the tinctures for proving.

Hahnemann did not leave us any details of the doses he used or the manner of giving the drugs, but from chance remarks elsewhere in his writings and from the accounts of his provers we have a pretty fair idea of what went on.

All the provings at this time were carried out with mother tinctures (extracts) of herbs or, in the case of insoluble substances, with "first triturations" (one part of substance ground up with nine parts of sugar of milk). That is, Hahnemann used actual material doses for the provings. I emphasise the point because it is often believed by homeopaths that he used high dilutions ("potencies"). In fact, he did not do so until much later.

His usual practice seems to have been to give repeated doses until some effect was produced; the actual amount was calculated on the basis of his own previous experience. The provers were expected to record their symptoms with the utmost care, and on presenting their notebooks to Hahnemann they had to offer him their hands — the customary way of taking an oath at German universities at that time — and swear that what they had reported was the truth. Hahnemann would then question them closely about their symptoms to elicit the details of time, factors that made them better or worse, and so on. Coffee, tea, wine, brandy and spices were forbidden to provers and so was chess (which Hahnemann considered too exciting), but beer was allowed and moderate exercise was encouraged.

The results of the Leipzig provings were published between 1811 and 1821 in a major six-volume work usually referred to as *The Materia Medica Pura*. As he had done earlier, Hahnemann supplemented his researches with reports of poisoning and over-dosage, and the resulting compilation was a unique contribution to pharmacology; nothing like it had been attempted before, and the information it contains (together with that in *The Chronic Diseases*, which I shall discuss later) still forms the basis of homeopathic practice today.

Not many modern homeopaths, however, make use of *The Materia Medica Pura* directly; instead they rely on secondary or tertiary sources. This is because Hahnemann unfortunately chose to present his findings in a way that makes them virtually unreadable. Instead of giving narrative descriptions of the provers' experiences he recorded their symptoms in an anatomical scheme of his own devising, so that what we are left with is a series of disconnected snippets that cannot be put together in the mind to yield a whole picture. As the nineteenth-century homeopath Robert Dudgeon remarked, it is as if a portrait gallery of family pictures were arranged by features — all the noses in one place, all the eyes in another, and so on. For this reason Hahnemann's original provings are seldom referred to today.

A further problem from our point of view is that Hahnemann's method of conducting his provings, though extremely meticulous and painstaking, did nothing to eliminate the effect of suggestion. The subjects knew what medicines they were taking (indeed, they had often gathered the herbs themselves) and they therefore knew what effects they might experience.

It is unfair to critici e Hahnemann for not recognising the importance of suggestion, for this was not properly understood until many years later, yet it has to be kept in mind in assessing his findings. Another difficulty with the provings is that all the provers were men (although it is likely that Hahnemann had earlier tried the medicines on female members of his family). But in spite of any reservations one may have there is no doubt that Hahnemann's Leipzig provings are a fascinating piece of work and represent a serious scientific attempt to investigate the properties of drugs.

It would be reasonable to expect that this achievement would represent the summit of Hahnemann's career and that he would now remain in Leipzig, surrounded by his small but devoted band of followers, while his own fame and that of his system spread ever farther and won new converts. After all, he was now in his sixties and he had made a name for himself professionally; it was hardly likely that he would now contribute any new ideas. And yet, much still lay in the future.

Departure to Köthen

Hahnemann's very success made him the target of much hostility, not only from doctors but also from his old enemies, the apothecaries, who resented the fact that he made up his own medicines and advised his disciples to do likewise. For a time their criticisms were silenced by the arrival in Leipzig of the victorious Prince Schwarzenberg, the hero of the battle of Leipzig, who came for the express purpose of being treated by Hahnemann.

Unfortunately, after an initial improvement the Prince died, and there was no lack of voices to accuse Hahnemann of having precipitated his demise. The apothecaries now obtained an injunction to prevent Hahnemann from dispensing his own medicines, and since they were unwilling to keep them themselves his practice could not continue. He was therefore forced to leave Leipzig.

The Duke of Anhalt Köthen, a small principality some 36 miles away, was an ardent admirer of the new system, and he offered Hahnemann the post of court physician in the tiny capital of his dominions. Hahnemann had no choice but to accept.

The move to Köthen took place in 1821. A considerable change came over Hahnemann in his new home. He was now virtually cut off, not merely from mainstream medicine but even from his own disciples. He became in effect a reclusive, hardly venturing outside his house. But he was by no means inactive; patients suffering from various forms of chronic disease came to him from all over Europe, and he continued to think, write and develop his system, which now began to take on new characteristics.

While he was in Köthen he published a third, fourth and fifth edition of *The Organon*, and also a second and third edition of *The Materia Medica Pura*. It was in Köthen, too, that he elaborated his famous notion of dynamisation and also announced his theory of chronic disease, both of which I shall discuss in the next chapter.

The hostility that homeopathy evoked from orthodox physicians and from apothecaries is easy to understand, but matters were undoubtedly made worse by Hahnemann himself. It may be the case that, had he not been so eccentric and obstinate, he would not have thought of homeopathy in the first place or have had the determination to defend and propagate his ideas in the teeth of opposition. But this independence and prickliness were to create needless difficulties for the new movement, which took on many of the attributes of a religious sect.

As so commonly happens in such sects the most virulent controversy occurred, not with outside critics, but within the ranks of homeopathy itself. For much of this dissent Hahnemann was himself responsible. From his seclusion in Köthen he continued to cause confusion in Leipzig. He dissolved the newly formed Homeopathic Society in Leipzig on the grounds that some of its members were not fully committed to the new doctrine, and his intolerance for deviation eventually became so extreme that he used to say: "He who does not walk exactly the same line with me, who diverges, if it be but the breadth of a straw, to the right or the left, is an apostate and a traitor and I will have nothing to do with him."

Soon after Hahnemann's departure a homeopathic hospital was established at Leipzig by private subscription and a Dr Müller was put in charge and gave his services for nothing. But Hahnemann took exception to Müller for his independence, and had him replaced by a salaried director. This man in turn was replaced by a bogus homeopath appropriately named Fickel, who took the job with the intention of discrediting homeopathy, and the consequent fiasco led in 1842 to the closure of the hospital.

As the years went by and Hahnemann aged he grew increasingly out of touch with general medical thought, but this did not prevent him from engaging in acrimonious disputes with the most eminent medical authorities, whom he treated with undisguised contempt. It has to be said that his arguments were by this time almost invariably superficial and irrelevant, for he was so utterly convinced of his own rightness that any attack, however well reasoned, seemed to him an expression of pure prejudice and ignorance.

Second marriage

In 1830, when he was 75, Hahnemann's wife died. They had been married for nearly 48 years and had had eleven children. Now, surely, Hahnemann's long life and career were all but over? But the last, and in some ways most remarkable, episode was still to come.

In October 1834 a mysterious visitor arrived at Köthen: a smart young Frenchman, whom the customary visit of the barber next morning unmasked as a beautiful girl. Mademoiselle Marie Melanie d'Hervilly, as the young lady was named, gave out that she had come to consult Hahnemann about her health.

However, a good deal of mystification attends both Melanie and the circumstances of her visit. She was about 32 to 35 years old at the time (she kept her exact age a secret). She had had a happy childhood in Paris but, according to her own account, her mother became jealous of her as she grew older and so she was adopted by a Monsieur and Madame le Thière.

Later she became well known as a portraitist and this gained her the entrée into the best social and intellectual circles, in which she had many influential friends. She seems to have been something of a feminist and to have felt strongly about the restrictions imposed on women by society; she had always had a leaning towards medicine, but of course at that time it was out of the question for her to study it.

In explanation of her visit to Köthen she said that her health had suffered owing to grief caused by the loss of several friends. She read *The Organon* and resolved there and then to visit its author. Not much is known about what happened next. What is certain, however, is that within three months of her arrival in Köthen — in January 1835 — Melanie and Hahnemann were married.

This event caused widespread astonishment. Hahnemann's numerous enemies naturally used the occasion to mock him, while his unmarried daughters, who kept house for him, were understandably less than enthusiastic; but Hahnemann himself found the experience reinvigorating and rejuvenating.

Six years earlier he had declined an invitation to visit Naumberg, on the grounds that travel had become impossible for him so that he could not even visit his married children. Three months after his marriage, however, Melanie took her husband off to Paris, leaving Hahnemann's two unmarried daughters to live out their lives in virtual seclusion.

Homeopathy was already established in Paris and Hahnemann was made welcome there. It was expected that the Master would restrict his activities to writing, but instead he took up medical practice and soon became very successful. In the vigour of his Indian summer he even went so far as to reverse his long-established custom of not making home visits and would drive out to patients and pay house calls even up to midnight.

Melanie assisted him, studied homeopathy under his tuition, and became a practitioner herself. The prosperous couple acquired a large house in the Rue de Milan, and Hahnemann, who had always been accustomed to living simply and frugally, now found himself in circumstances that were comfortable, even luxurious. There seems no doubt that his final years with Melanie were happy, and though many of his followers, both during his lifetime and later, attacked her bitterly, Hahnemann himself apparently found peace and fulfilment with her.

Hahnemann died on 2nd July 1843. Melanie kept the funeral private, and his biographer Haehl implies that she forgot him as soon as he was buried; but this seems at variance with the fact that when Hahnemann's body was disinterred in 1896 a lock of Melanie's hair was found round his neck.

Disagreement among Hahnemann's followers by no means ceased at his death. Much of this concerned the Master's literary relics, including the sixth edition of *The Organon*, on which he had been working shortly before his death. This material remained in the possession of his widow, who continued to practise homeopathy. At her death it passed to her adopted daughter, who had married the son of von Boenninghausen, one of Hahnemann's most devoted disciples. After many difficulties Haehl succeeded in obtaining the manuscript, which was finally published in 1922.

Hahnemann's life and main publications

1755   Born at Meissen

1779   Qualifies in medicine at Erlangen

1782   First marriage

1782-1805   Years of wandering

1790   Cinchona experiment

1806   Publishes *Medicine of Experience*

1810   Publishes first edition of *The Organon*

1811   Settles in Leipzig. Provings and publication of *The Materia Medica Pura*

1821   Moves to Köthen. Publication of *The Chronic Diseases*

1830   Death of first wife

1835   Marriage to Melanie. Moves to Paris

1843   Death in Paris.

Chapter 3

Hahnemann's Later Ideas

As Hahnemann aged he began to take homeopathy in new directions. He introduced an explanation for how the medicines work based on vitalism, he invented the idea of potentisation, and he came up with a theory of chronic disease, which he introduced to account for failures of treatment in patients suffering from these diseases.

Vitalism

In the later editions of *The Organon* and also in his other writings of this period we find an increasing emphasis on the doctrine of vitalism. The term Hahnemann used was *dynamis*, which is usually translated as "vital force". By this he meant a spirit-like principle that gives life to the body. Disease, he came to believe, results from disturbances in the vital force produced by outside influences of various kinds, and the function of homeopathic medicines is said to be to stimulate the vital force to bring about healing.

Hahnemann did not of course invent the idea of the vital force; in one form or another it is probably as old as humanity. It appears to be an almost universal primitive belief that there is such an animating spirit in man, often identified with the breath (*pneuma* in ancient Greece and the writings of St Paul, *prana* in India), which leaves the body at death and is responsible for its functioning during life. Plato presents a sophisticated philosophical version of this idea and it can be traced in Western philosophy down to modern times (for example, in the writings of Henri Bergson), though it is outlawed in mainstream science today.

In Hahnemann's time vitalism was still a serious scientific idea. At the beginning of the eighteenth century Ernst Georg Stahl had taught a form of vitalism and his ideas continued to be influential among doctors in Germany and France, particularly at the University of Montpellier. The true nature of the life force was held to be unknown and unfathomable. It had its seat in the brain and solar plexus and transmitted its influence via the nerves, believed to be hollow. Disease was supposed to be due to disturbance of this force and healing took place through its operation, though the assistance of the physician might be needed at times.

These ideas were advocated by Hufeland, the editor of the journal in which many of Hahnemann's early essays on homeopathy appeared. It is therefore not surprising that Hahnemann adopted vitalism as a basis for homeopathy, though it was only in the later editions of *The Organon* that he did so. (In the early editions he was if anything dismissive of the idea.)

Hahnemann's increasing sympathy for vitalism is symptomatic of a general shift in the centre of gravity of his thought, from what might be called the scientific to the spiritual or metaphysical pole. It is possible to discern two phases in his development, in which first one tendency and then the other predominated. Although the division between the two periods is not absolute we can say that the watershed was the year 1821, in which he left Leipzig for Köthen.

Up to this time he was on the whole a scientist, carrying out his provings, modifying his practice in the light of experience, and associating with other doctors. In his seclusion at Köthen he continued to speculate and to change his ideas, but in directions that led him further and further from mainstream medicine. Because he was cut off even from his own followers he was practising and thinking in a vacuum, and his ideas became ever more extreme. It is mainly from this period that derive those features that have tended to isolate homeopathy from orthodox medicine.

I underline this distinction between the two phases of Hahnemann's career because it seems to me to explain much of the later development of homeopathy. On the whole, homeopaths after Hahnemann were led by their temperaments to emphasise one aspect of his thought to the virtual exclusion of the other. There have been those who have laid more weight on Hahnemann's scientific characteristics and have regretted the vitalistic ideas in *The Organon*, and there have been others who have on the contrary magnified the differences that separate homeopathy from mainstream medicine. In a sense, the rest of this book will be concerned with the results of this difference of opinion.

In addition to vitalism, Hahnemann introduced into homeopathy two other new ideas during his sojourn in Köthen: the potency doctrine and the theory of chronic disease. So important are these two dogmas (for that is what they became) for the subsequent development of homeopathy that we need to take a little time to examine them.

The potency idea

When Hahnemann first thought of homeopathy he used big doses, like the orthodox physicians of his day. Quite soon, however, he switched to using very small doses. He did this to reduce the unwanted effects of the medicines; there was no question at this stage of making the medicines *more* effective. On the contrary, diluting the medicines did weaken them, he said, but not nearly as much as might be expected. In any case, he claimed, when people are ill they become abnormally sensitive to medicines and so need smaller doses.

So matters stood in the early part of Hahnemann's homeopathic career. By 1825, however, when he was at Köthen, he had adopted a radically new idea: dynamisation. This emerges from an answer he gave to a critic who said that to use homeopathic doses was like putting a drop of a drug in Lake Geneva and using the water for medicine. Hahnemann rejected this comparison on the grounds that the method used to prepare homeopathic medicines was not a mere dilution but involved dynamisation or trituration, which released astonishing powers in them; active substances were made more active and hitherto inactive ones, such as quartz sand, were found to have unsuspected latent properties. So the potency effect was not part of homeopathy as originally described, but it has always been the feature that has most strongly attracted the scorn of critics.

Hahnemann tried to explain dynamisation by comparing it to the production of heat by friction and to magnetisation by stroking a piece of steel with a magnet, neither of which was understood in his day. Dynamisation was for Hahnemann a process of releasing an energy that he regarded as essentially immaterial and spiritual. As time went on he became more and more impressed with the power of the technique he had discovered and he issued dire warnings about the perils of dynamising medicines too much. This might have serious or even fatal consequences, and he advised homeopaths not to carry medicines about in their waistcoat pockets lest they inadvertently make them too powerful. Eventually he even claimed that there was no need for patients to swallow the medicines at all; it was enough if they merely smelt them. Few of his followers, however, were prepared to go as far as this. Indeed the whole potency idea was difficult for some homeopaths to accept and it was to become a fruitful source of controversy in later years.

Superficially, perhaps, the potency concept might seem to be scientific. Hahnemann certainly claimed that the superior effectiveness of potentised medicines had been amply demonstrated in practice. Yet from his writings it is evident that his reasons for adopting the theory had other roots. What really appealed to him about it was its connection with the idea of the vital force. Potentised medicines were for him the vital force captured in a bottle. And, as with all his later innovations, once he had thought of potency it became an integral and essential part of homeopathic theory and he had no time for any of his followers who expressed any doubt about it.

Potency today

The potency idea is undoubtedly the aspect of homeopathy that has most strongly captured public attention. People who know nothing else about the subject usually are at least aware that homeopaths use medicines in tiny doses, and critics often quote this to show that homeopathy is self-evidently absurd. It is pretty well established as integral to homeopathy today.

The modern position is as follows. Nearly all homeopathic medicines are made by a process of alternate dilution and "succussion" (violent shaking). The succussion is an essential part of the

Decimal (x}	Centesimal (c)	Dilution
1x	—	1:10 ($10^{-1)}$)
2x	1c	1:100 (10^{-2})
3x	—	1:1000 (10^{-3})
4x	2c	1:10,000 (10^{-4})
5x	-	1:100,000 (10^{-5})
6x	3c	1:1,000,000 (10^{-6})
(etc}	(etc}	(etc}
12x	6c	1:1,000,000,000,000 (10^{-12})
24x	12c	10^{-24}

Table 3.1: Potency Scales

procedure, though now it is usually done mechanically instead of by hand, which was Hahnemann's practice. (He used a Bible as a succussion pad.) Succussing the medicines is supposed to increase their activity and this is what distinguishes a homeopathic medicine from an ordinary solution. Increasing the effectiveness of a medicine in this way is referred to as potentisation or dynamisation — the terms are interchangeable — and the medicines are commonly called "potencies".

Two potency scales are in common use: the decimal, which proceeds by 1:10 steps, and the centesimal (1:100). (Towards the end of his life Hahnemann invented another method of potentising, the "LM scale", but it has not been widely used by later homeopaths so I will ignore it here.) Starting from the original "mother tincture" (in the case of a plant this is an alcoholic extract) a 1:10 or 1:100 dilution is made, by taking either 9 or 99 drops of solvent and adding one drop of tincture. The mixture is succussed and the resulting solution is known as the first potency. This now serves as the starting point for the next step in dilution and succussion, which results in the second potency, and so on. The 1:10 potencies are usually indicated by x and the 1:100 by c; thus pulsatilla 6c means the 6th centesimal potency of pulsatilla, which has received six succussions and has a concentration of one part in a million million or 10^{-12} (Table 3.1).

Insoluble substances, such as metals, are prepared by grinding them together with lactose (milk sugar) in the same 1:10 or 1:100

proportions. This process is called trituration and is supposed to be equivalent to succussion. After the 6th trituration the particles become so finely divided that they can form colloidal solutions in water, and then liquid potentisation continues in the usual way.

A 30c potency, which is widely used, is a 10^{-60} dilution! Allegedly much higher potencies are also made; I return to this later. In Britain the 6c and 30c potencies are generally used; the intermediate potencies (4c, 5c, 9c etc.) are not available except by special prescription. Potencies of this kind are however widely used in France. For reasons I discuss in Chapter 5, 24x (12c) is taken as the boundary between "molecular" and "ultra-molecular" potencies.

The theory of chronic disease

According to Hahnemann himself, he first devised his chronic disease theory in the years 1816-17 — that is, while he was still at Leipzig — though he did not make it public for a further decade. In 1827 he summoned his two closest followers to Köthen to receive the new doctrine, and in the following year he began to publish his last major work, *The Chronic Diseases*, in which his theory was set forth. The new book eventually went into a second edition; nevertheless it did not sell well and the theory itself was contentious.

Hahnemann was led to formulate his theory by the discovery that although homeopathy appeared to be effective enough in the treatment of acute disease many difficulties were encountered in the treatment of chronic disease. Patients often seemed to respond to the medicine initially, but later they ceased to do so or produced new symptoms in place of the old. Some homeopaths supposed that the answer would come from the proving of new medicines, but Hahnemann rejected this solution and instead produced his own answer: the miasm doctrine.

In outline the theory can be stated quite simply: all chronic disease, apart from that due to orthodox medicine or to faulty living habits, is caused by one of three "miasms" — syphilis, sycosis, and psora.

Hahnemann did not invent the term miasm, which was already in use in orthodox medicine in his day, but he gave it a new meaning and scope. The word derives from the Greek and means something like "taint" or "contamination". Hahnemann supposed that chronic disease results from invasion of the body by one of the miasms through the skin. The first sign of disease is thus always a skin disorder of some kind. This may clear up, either spontaneously or — much worse — as the result of allopathic treatment, but the miasm will infallibly have spread throughout the body and will give rise to all kinds of problems in later years.

To a modern reader this description suggests almost irresistibly the notion that the miasms are infections. Hahnemann did actually toy with the idea of microbes in another context, for he suggested in the case of an acute disease, cholera, that it might be caused by a minute organism too small to be seen. However he does not seem to have made the same suggestion about the chronic miasms. Nevertheless the temptation to call them infections is almost overwhelming.

What is particularly interesting about Hahnemann's theory is that in the case of syphilis he was more or less right. We now know that syphilis is caused by an infection that enters via the skin, producing an apparently localised disease — the chancre. From the beginning, however, the infection is generalised, and if untreated it does go on to cause all kinds of serious and even fatal effects. Syphilis is therefore a good example of a miasm.

The typical lesion of Hahnemann's other venereal miasm, sycosis, is fig-warts (genital warts). However, any kind of warty growth anywhere on the body is supposed to be sycotic and so are discharges of various kinds. Sycosis includes what we would now call gonorrhoea but it is much wider in scope.

So much for the two venereal miasms. The third chronic miasm, psora, is much more important than both of the venereal miasms put together, for it accounts for seven-eighths of all chronic disease not caused by faulty patterns of living or by allopathic medicine. The skin manifestation of psora is typically scabies (the itch). Today we regard this as due to a mite that burrows in the skin, but Hahnemann's conception of psora is much wider than this and almost any kind of non-warty skin eruption, especially if itchy, is supposed to be psoric.

The course of psora is very similar to that of syphilis. First the patient suffers a skin disease, which may be so trivial or have happened so long ago that he has forgotten it. There then follows a latent period lasting months or years during which there are few or no symptoms, until at last the psora breaks forth in any of the innumerable forms of chronic disease.

Psora is extraordinarily infectious. It can be passed on, especially to children, simply by touching the skin. A mother can give it to her baby during delivery, a doctor can transmit it by feeling the pulse, or it can be carried in clothing or bedding. So infectious is it that scarcely anybody escapes; in fact the only mortal fortunate enough to have done so appears to have been Hahnemann himself, for he solemnly assures us that it is thanks to his unique freedom from the psoric trait that he has been able to detect it in others.

Assessment of the miasm theory

What are we to make of this remarkable theory? The important point, I think, is that it is not what it seems. Superficially it appears to be a pathological scientific hypothesis about the mechanism of disease. This in itself was an implied contradiction, for it was a cornerstone of Hahnemann's system that nothing could be known about the underlying mechanisms of disease; hence he could be — and was — accused of inconsistency in advancing a pathological theory. Inconsistency never troubled Hahnemann but even so it is at first glance rather surprising to find him advocating an idea of this kind.

In fact, however, the miasm theory, though it masquerades as a pathological theory, is really nothing of the kind. A genuinely scientific theory ought to be open to being tested in some way, but there is no conceivable way to test the miasm theory as Hahnemann presents it. In *The Chronic Diseases* Hahnemann gives a most extraordinary list of symptoms that are supposed to be due to psora. It takes up some 33 pages and includes almost every ill known to human beings — and even so Hahnemann tells us that it is incomplete. But if every imaginable manifestation of chronic disease is due to psora, how does the theory help us? A theory that tries to explain everything really explains nothing. (To see what I mean, read Hahnemann's description and then try to think of a disease or symptom that would *not* be due to psora.)

The only conclusion we can draw, I believe, is that the miasm theory was a face-saver. It was introduced by Hahnemann to preserve the inviolability of his system. He had been forced to acknowledge that homeopathy was not universally successful but he could not admit the thought that it was not a complete answer to disease, since he had invested too much of himself in it psychologically. The only way out of the impasse he could find was to postulate the existence of a deep-seated almost ineradicable hydra-headed evil.

But not quite ineradicable, of course. For the elimination of the psora monster Hahnemann described a group of new "antipsoric" medicines. The chief of these was sulphur, which was already in use at the time to treat scabies, though by external application, something Hahnemann strongly disapproved of. But there were also several others, including some very unlikely-sounding substances — for example, sepia (cuttlefish ink), natrum muriaticum (common salt), and silicea (quartz sand). By the judicious use of these medicines it would usually be possible to eradicate psora but the process might take several months or even years, and if the infection had been previously "driven inwards" by unwise application of external medicines to the skin, cure might be totally impossible.

The medicines introduced by Hahnemann in *The Chronic Diseases* were destined to become very important in homeopathy. However it seems almost certain that they had not been "proved", at least by Hahnemann, in the accepted manner — that is, by experiments on healthy volunteers. They hardly could have been, for Hahnemann was by now too old to carry out provings on himself and he was living in almost complete isolation from his colleagues.

What appears to have happened is that Hahnemann based his new provings largely on symptoms supposed to have been produced *in his chronic patients*. By his own rules this procedure was inadmissible, and in fact it undoubtedly led him to attribute to the effect of the medicines a number of symptoms that were really due to the diseases the patients were suffering from. Moreover they were also, apparently, obtained with 30c potencies instead of the material doses used by Hahnemann in his earlier provings at Leipzig. It is questionable whether 30c provings are capable of causing symptoms. For these reasons critical homeopaths, such as the nineteenth-century

English homeopath Richard Hughes, have been suspicious of the symptoms of medicines recorded in *The Chronic Diseases*.

Whatever one's opinion of the scientific status of the psora theory as put forward by Hahnemann may be, there is no denying that the idea became increasingly remote from science in the hands of many of his successors. For Hahnemann the miasms were acquired "infections"; people were not born with them but suffered them (in the case of psora) at or soon after birth. Oddly enough, Hahnemann does not even seem to have recognised the existence of congenital syphilis. In principle it was possible to avoid infection altogether, as Hahnemann himself was fortunate enough to have done. By a curious historical reversal, however, many later homeopaths have praised him for his supposed recognition of the *hereditary* element in chronic disease. The explanation of this lies in the way that homeopathy developed in the USA, which will be my subject in a later chapter.

Hahnemann's religious outlook

The real importance of the miasm theory, it seems to me, is the insight it gives into how Hahnemann thought about homeopathy. We shall not understand the man unless we realise that, for him, homeopathy was much more than a mere medical theory; it was a divine revelation.

I am not exaggerating here. We know from his own writings that the idea of homeopathy came to him as the solution to a religious dilemma. This dilemma was the paradox that confronts anyone who believes in a God who is simultaneously all-powerful and all-good: how to account for suffering? Hahnemann was not a Christian but he was a deist. He believed that the universe had been designed by an infinitely wise and loving Father, and such a Father, he reasoned, must have provided his children with a means of relieving their suffering. But what?

At first he could see no solution. As late as 1805, the year before the publication of *The Medicine of Experience*, we find him writing almost in despair:

> After 1000 to 2000 years, then we are no further! How turbid art Thou, sole source of our knowledge of the powers of medicine! And yet in this cultured century this state of affairs is perfectly satisfactory to the learned bevy of physicians, in the most important affairs of mortals, where the most precious of all earthly possessions — human life and health — are at stake!

The problem continued to obsess him as the years went by. In 1808 we find him still writing on the same theme, though by this time he had already discerned the divine Answer to the enigma. After a lengthy description of his progressive disillusionment with orthodox medicine Hahnemann explains that he was at last driven to wonder whether "perhaps the whole nature of this science, as great men have already said, is such that it is not capable of any great certainty". No sooner does he consider this shocking idea, however, than he rejects it decisively.

> What a shameful blasphemous thought! — I clasped my brow — that the wisdom of the Infinite Spirit animating the universe would not be able to create means to relieve the sufferings of diseases which He, after all, allowed to arise ...
>
> Would He, the Father of all, coldly survey the torments of disease of His dearest creatures? Would He leave no way open to the genius of mankind — otherwise so infallible — no easy, certain and dependable way of regarding disease from the right angle, of determining the use and the specific, safe and dependable results obtainable from the medicines? Before I would have given credence to this blasphemy I should have forsworn all the school systems of the world . . .

There was thus a deeply religious element in Hahnemann's conception of homeopathy right from the beginning and as time went on this came to predominate more and more, which helps to explain why he eventually regarded anyone who criticised him almost as a blasphemer and any disciple who deviated from his line of thought as a renegade. We have already seen the unfortunate effects that this inflexibility produced on the homeopaths of Leipzig.

What all this amounts to is that Hahnemann took on the mantle of a guru. In his book *Feet of Clay* the psychiatrist Anthony Storr has written of the phenomenon of gurudom. "Guru" is a Sanskrit word and in its original sense it refers to an Indian religious preceptor, who is often invested with quasi-divine authority by his (occasionally her) disciples. By extension the term is now sometimes applied to Western authority figures. Storr lists a number of psychological characteristics that people we would class as gurus commonly exhibit.

Gurus commonly have an isolated childhood. We know little about Hahnemann's childhood, principally because he himself had little to say about it; the suggestion is that it was not particularly happy. Whether or not this was the case, there is no doubt that Hahnemann's early medical career was a time of stress and unhappiness; he was in straitened circumstances financially and he was disillusioned with conventional medicine. Storr finds that gurus generally experience a period of psychological distress before beginning their career as teachers and public figures. This period of unhappiness typically comes to an end when the guru receives a revelation, which often has an explicitly divine origin even when its content is mainly secular. This is certainly true in Hahnemann's case.

Hahnemann's discovery of homeopathy had for him the quality of a religious revelation, and this awareness of divine guidance and inspiration never left him. When he was on his deathbed in Paris his wife Melanie spoke of how much his patients owed to him for his life-saving discovery, but Hahnemann disclaimed all credit for it, saying that thanks were due, not to him, but to the Father.

It is in this context that we must place the extreme sensitivity which Hahnemann, like other gurus, felt towards any form of criticism from outside or questioning from his own disciples. As Storr remarks, it is characteristic of gurus to react in this way. Any disagreement with the guru is interpreted as hostility, and those who express such disagreement are liable to be greeted with abuse. Eventually it became almost impossible for even Hahnemann's most loyal disciples to remain on good terms. One of these had the misfortune to lose a child, and wrote to Hahnemann to say that his loss had taught him that homeopathy did not suffice in every case.

Hahnemann was so incensed at this that he never forgave the man or restored him fully to favour.

Newcomers to homeopathy are often struck by the frequency with which Hahnemann's name is invoked in talks and articles. *The Organon* is still often quoted as if it were Holy Writ. I can think of no other example of a scientific or medical author since Galen whose authority has lasted anything like as long as Hahnemann's, though admittedly only within the small circle of homeopaths, for outside homeopathy his name is almost unknown.

For better or worse, Hahnemann is an excellent example of a guru. He is however by no means the only homeopathic guru, even if he is the principal one. The subject seems to have an irresistible attraction for people who are temperamentally inclined to gurudom. One of the chief examples of this is the American nineteenth-century homeopath James Tyler Kent, whom I discuss at some length in a later chapter, but there have been many others right down to our own day.

Such people make confident *ex cathedra* statements, nearly always without quoting any real supporting evidence, and they tend to attract around them bands of admiring disciples who propagate their Masters' ideas with missionary zeal. Homeopathic gurus follow the example of Lewis Carroll's Bellman: "'I have said it three times,' said the Bellman / 'And what I say three times is true.'"

Chapter 4

The Years of Hope

At Hahnemann's death his doctrine had already spread widely in spite of opposition, and homeopaths felt confident that it would not be long before they had achieved recognition as practitioners of the one rational form of medicine. Homeopathy was to be found not only in Germany and France, the two countries where Hahnemann had practised, but also in England, Italy, Spain, Scandinavia, Poland and Russia; it had crossed the Atlantic to both Americas and it had taken root in India, still the country where it flourishes most successfully

At this time homeopaths were not content to rest upon the labours of the Master; many of them took up various aspects of his teaching and developed them in new directions. In the next chapter I shall look at what happened to the potency idea but here I am concerned with the new provings that were undertaken by some of his more adventurous disciples, sometimes at considerable personal risk.

In spite of their fundamental importance for homeopathy, both of Hahnemann's major contributions to pharmacology — *The Materia Medica Pura* and *The Chronic Diseases* — had serious flaws. *The Materia Medica Pura* was so arranged as to make it almost unreadable, consisting as it does of mere lists of symptoms arranged anatomically, so that, in the words of Richard Hughes, an eminent British homeopath of the day, the would-be reader of Hahnemann's article on Aconite begins with Vertigo and ends with Rage. As for *The Chronic Diseases*, it suffered from the same problems of arrangement and in addition, as we saw in the last chapter, there were doubts about its reliability.

Many homeopaths, therefore, while not questioning Hahnemann's genius or importance, felt that there was a need to re-prove his medicines to see whether the symptoms he had found could be reproduced. They also wanted to test new medicines for their possible application to disease. For these reasons the second half of the nineteenth century saw an astonishing spate of provings, especially in Germany, Austria, and the USA. Some of the most interesting and extensive provings of the later nineteenth century were carried out by the Austrian Homeopathic Society, which re-proved a number of Hahnemann's medicines and also some new ones.

The narrative accounts of the nineteenth-century provings are often dramatic. The provers often went to lengths that can only be called heroic, and their records provide striking evidence of their homeopathic zeal. It is certain that work of this kind will never be repeated, which makes it of unique historical interest if nothing else. And yet the surprising fact is that today all these original reports remain locked away in nineteenth-century tomes, gathering dust and almost unread even by homeopaths. Later we shall see how this has come about but for the present let us look at what happened to some of these pioneers.

The medicines that were tested fall into three broad categories. First, there are substances that are definitely poisonous if taken in adequate dosage — mercury, phosphorus, and arsenic, for example. As might be expected, provers who took these substances often made themselves quite seriously ill. Second, there are substances that, although certainly capable of making people ill, seldom cause death even when taken in fairly large doses. In this second group we find, for example, nutmeg, hashish, and poison ivy. Third, there is a group of substances that would ordinarily be thought of as more or less inert or harmless; here we find common salt, charcoal, and quartz sand. This group offers special difficulties to a modern reader, in that it is particularly difficult to decide how far the symptoms attributed to the medicines may really be due to something else. There is a certain amount of overlapping among these three categories but in what follows I shall treat them separately.

Dangerous poisons

The nineteenth-century homeopathic literature contains many alarming reports of people taking appallingly large doses of poisonous substances. Constantine Hering, for example, took a lead preparation until his moustache and eyebrows fell out and his teeth decayed. A Dr Spence also took lead in increasing doses over three weeks; his gums became spongy and he suffered other well-known symptoms of lead poisoning such as colicky abdominal pain and paralyses of his limbs. Knowing what we do today about the persistence of lead in the body and its long-term effects, we must assume that these provers would have continued to suffer from lead poisoning long after the end of the experiments.

Another poison that attracted a great deal of attention from provers was phosphorus. This was used at the time in the manufacture of matches and was well known as an industrial poison; workers in the industry suffered loss of teeth and destruction of their jaw bones ("phossy jaw"). Provers who took phosphorus duly suffered pains in their teeth and facial bones. Some of them also experienced interesting psychological symptoms. Dr Heath, an American, took five drops of phosphorus tincture and then dismissed the matter from his mind. At about 10 p.m. he went to bed but was unable to sleep.

> My mind was greatly oppressed with melancholy; tears would start without cause; a feeling of dread, as if awaiting something terrible while unable to resist or move, overcame me. Sometimes it seemed as if I were beginning to bloat, and then I could hear a multitude of voices saying, in high glee: "Fill him up a little more and he will burst," followed by demoniacal laughter ... When I attempted to walk my legs seemed glued to the floor; the slightest motion caused great pain ...

It took Heath more than two months to recover fully from this alarming experience.

Sometimes provers took matters to the point where their health was permanently impaired. This occurred during the provings of arsenic in America, in which some of the provers took doses of a tincture of arsenic for long periods — fifty days or more. They experienced a variety of symptoms, some of which lasted for over two years.

Toxic but non-fatal substances

This group contains some of the most interesting accounts, since the provers often took large doses over many months. This comes out particularly clearly in the case of Thuja occidentalis, the Tree of Life. One hardy experimenter took 42,260 drops of tincture over 155 days; some others took nearly as much. Not surprisingly an enormous variety of symptoms ensued, which it is quite impossible to summarise. Thuja was for Hahnemann the principal anti-sycotic (anti-wart) medicine, and in fact a number of provers, including three children, developed warts (the apparent willingness of some enthusiasts to experiment on children and even babies is remarkable), and in adults a gonorrhoea-like urethritis was also seen. Dr Robert Dudgeon, a prominent English homeopath of the time, had in this connection an embarrassing experience which I give in his own words.

> On 10th July, when taking a walk, I happened to pass a [Tree of Life] laden with green cones. I plucked one, chewed it a little, and thought no more about it. That same evening I observed a very disagreeable scalding on making water, which continued all next day; and I was horrified to observe on undressing that my shirt was spotted all over in a manner extremely repugnant to one's notion of respectability ... I had quite forgotten the circumstance of having chewed the Thuja cone and could not imagine what could have produced in me, a decent paterfamilias, such a very incongruous complaint. The following day the discharge had become yellow ... I now remembered the cone-chewing and regarded the malady with more composure. [The disease lasted until the 16th.] The symptoms ... were precisely those of an ordinary attack of gonorrhoea, but their medicinal origin was evidenced by the short duration of the attack.

> I should add that two colleagues who, at my suggestion,
> chewed a cone as I had done were unaffected by it.

There are no reports of fatalities from accidental overdosage with
thuja, and from the large doses taken by some of the provers it would
not seem to be a dangerous substance. However, the symptoms often
lasted a long time, sometimes for a month or more after the last dose
was taken.

Another interesting substance is poison ivy (Rhus species). This
is a plant that grows wild in North America. People become sen-
sitised to it easily and then suffer severe skin reactions whenever
they come into contact with it. Hahnemann introduced the herb as a
homeopathic medicine in *The Materia Medica Pura* and it has always
remained an important item in the homeopathic pharmacopeia,
being used for the treatment of skin disorders and also certain kinds
of muscle and joint pains.

The American provers experimented with rhus quite extensively.
For the most part they used extracts of the leaves, either neat or in
low dilutions. Most of them experienced the expected skin and muscle symptoms but the details of some of the narratives are curious.
One prover, for example, became so exquisitely sensitive to the plant
that in subsequent years he was unable to pass a swamp in which
the plant was growing without suffering renewed symptoms. An
unusual feature was that at these times his wife would experience
vaginal burning after intercourse.

Another prover, a Dr Clary, held a stick of rhus in his hand for
half a minute and just touched his tongue with the tip of it. Nothing
happened for a week; then while sitting at dinner he suddenly felt a
scalding sensation in his tongue, and this grew rapidly worse and
spread over his whole mouth and throat. Over the next few days
he became very ill; a severe rash spread all over his body, his whole
intestinal tract was affected, and his muscles ached so much that he
could barely walk. It was more than two weeks before he recovered.

Other drugs in this group include those taken up as psychedelic
agents by later generations, such as nutmeg and hashish. Hallucina-
tions and other bizarre mental symptoms are reported surprisingly
seldom by provers, who seem mainly concerned with bowel disturbances, aches and pains, and various strange *physical* sensations.

(This is true of the proving literature taken as a whole.) But at least one prover experienced a bad trip after taking hashish.

> I felt myself mounting upwards, expanding, dilating, dissolving into the wide confines of space, overwhelmed by a horrible, rending, unutterable despair. Then, with tremendous effort, I seemed to shake this off, and to start up with the shuddering thought: "Next time you will not be able to throw this off, and what then?"

Apparently inert substances

This group is in some ways the most puzzling to evaluate. It is very difficult to understand how taking common salt or charcoal could produce genuine symptoms, yet these and similar apparently inert substances were extensively proved by the early homeopaths. According to Hahnemann they would not have had any effect in their crude form but must first be activated by dynamisation. Even homeopaths were sometimes sceptical about this, which makes their eventual conversion through personal experience all the more interesting. A good example is provided by the Austrian provings of common salt.

Some provers were in fact insensitive to salt. Others had well-marked symptoms and there are pages and pages about them. One of the most interesting reports is that of Dr Watzke, who on 2 March 1843 began to take salt in various doses. At first nothing much seemed to happen, but then he began to suffer pains in his joints of such severity that he could hardly walk. The symptoms continued until the end of May. Reflecting on his experience, Watzke wrote:

> It could not be easy for anyone to show themselves less susceptible to small as well as large doses of common salt than I showed myself at the beginning of my experiments ... And yet the medium doses, used continuously for a longer period, developed the salt disease in me almost to complete cachexia; and of all the medicines which I have hitherto proved, none created ultimately such a deep penetrating tenacious action in me as common salt.

This is certainly a curious account and it does not stand alone; numerous other provers reported something similar. Watzke appears to have been a sceptical and objective observer and it is difficult to dismiss his account as the result of error or self-deception.

This is not true of all the substances in this group. The provings of several of the major homeopathic medicines, such as silicea, sepia, and lycopodium are less convincing. A great many symptoms are attributed to these medicines but they are mostly rather indefinite and I at least am left with the suspicion that many of them are really due to suggestion or other factors.

A case in point is provided by a report of the supposed effect of musk. This substance, used in the manufacture of scent, is derived from a special gland possessed by the musk deer and was supposed to have aphrodisiac properties. It had been proved by Hahnemann and other researchers, but the most startling description of effects comes from Hromada and concerns the experiences of an unspecified number of people engaged in grinding (triturating) musk.

A man aged 52 spent an hour at the task. In the first five minutes he had "a kind of rush of blood to the head, with staring eyes and spasm in his mouth, so that he could not answer when asked what was the matter, though he understood what was said." He then began to speak, but rapidly and confusedly, and he would not stop when asked to do so. He became pale and sweaty and staggered as if drunk. His eyes rolled upwards, his jaw moved as if chewing, and he was unable to answer questions coherently. All these symptoms disappeared half an hour after he was taken into the fresh air.

Rather surprisingly, he then resumed his grinding duties. All went well for half an hour, but then the symptoms came back with greater force than before. He lost consciousness and suffered a hallucination of big black figures pressing in on him.

Other musk grinders had symptoms that were almost as alarming. One woman aged 45, for instance, lost consciousness, but before this happened everything seemed to round in a circle, at first slowly, then faster and faster until at last it seemed as if she were hovering in the air and then falling from a great height.

It seems surprising to say the least that grinding musk should have had such striking effects as these. What can have been going on? A clue, I think, is provided by the occurrence in one musk-grinder, a woman aged 60, of sexual desire. We are assured by Hromada that she had never in all her life had such a sensation before, but it is permissible to wonder whether she can have been quite so immune from the desires of the flesh as this.

It seems much more likely that many of the symptoms supposedly due to the musk were really caused by a combination of collective hysteria (assuming that all these people were doing their grinding together, which is implied though not stated) and suppressed sexual awareness heightened by associations to musk. The phenomena described by Hromada are remarkably similar to those that occurred in Anton Mesmer's groups. Mesmerism was fashionable at this time and there are numerous accounts of trances with a strong sexual element during Mesmeric sessions. (See Chapter 10.)

Support from this idea comes from an interesting case reported by Dudgeon.

> An unmarried lady of about 40 mentioned to me that she was extremely sensitive to the odour of musk. She would faint if she merely opened a note highly scented with musk. A doctor who was unaware of this peculiarity prescribed for her a pill containing $\frac{1}{4}$ grain of musk. Soon after taking this she became unconscious, was violently convulsed, and this state lasted nearly a week, with short intervals of consciousness. She said her life was despaired of.

In Chapter 11 I describe the results of a modern proving of pulsatilla, in which the participants produced so many symptoms in response to placebo that many of them withdrew from the trial.

Chapter 5

New Directions

When Hahnemann died he bequeathed to his disciples several difficult problems which have not been completely resolved even today. These related to three main areas.

First, there was the enormous and intractable volume of materia medica — the information about medicines derived from provings and from reports of poisoning and over-dosage in the general medical literature. Within Hahnemann's lifetime this was already massive and it continued to grow after his death. Somehow ways had to be found to make it more assimilable and easier to make use of in practice.

Second, there was the ever-troublesome question of potency. This idea had been hard enough to accept even in Hahnemann's day, but as the nineteenth century wore on it became more and more difficult to reconcile with scientific knowledge.

Third, there was the question how far, if at all, homeopathy could or should be related to the new medical ideas that were beginning to appear in the second half of the nineteenth century thanks to innovators such as Virchow, Pasteur and Koch, who were giving the germ theory of disease a scientific basis. Was homeopathy to remain aloof from orthodox medical ideas as Hahnemann had insisted or should it change with the times?

In this chapter I look at some of the ways in which homeopaths tried to solve these problems.

The Materia Medica

New recruits to homeopathy were understandably intimidated by the vast bulk of knowledge about medicines that they were expected to acquire. Moreover, some of this knowledge was not easy to get at. Hahnemann's writings were available, of course, but in addition there were many reports scattered about in homeopathic journals, and this literature was constantly growing as provings went on. The difficulty was compounded for homeopaths who did not read German. Attempts were therefore made to draw all the available information together into major reference works.

One of the earliest of these collections was prepared by a German called Jahr, but it had many shortcomings; the English homeopath Richard Hughes dismissed it contemptuously as "nonsense made difficult." In 1874 an American, T.F. Allen, began editing a new work, *The Encyclopaedia of Pure Materia Medica*. Allen included all the material he could find, without making any attempt to judge its reliability; the result was a daunting collection that ran to ten large volumes.

Such massive undertakings defeated their own ends. They were of no use to practitioners at the bedside, and indeed it is questionable to what extent they were used at all; certainly they gather dust today. Some homeopaths, especially in America, went to the opposite extreme and simplified the materia medica drastically, listing just the salient features ("keynotes") of each medicine in a couple of pages. This naturally appealed to newcomers to homeopathy but the purists scorned the "keynote method" as impossibly crude and simplistic.

Another development was the compilation of indexes to the materia medica. These "repertories", as they were called, were designed to allow practitioners to look up the medicines that corresponded to particular symptoms. Confronted with a feverish patient with a left-sided tonsillitis and a swollen right knee, for example, a homeopath could look up these symptoms in his repertory and see which medicines had corresponding symptoms. The best known of the early repertories was compiled by von Boenninghausen, a lawyer turned homeopath whose son married the adopted daughter of Hahnemann's second wife Melanie.

By the late nineteenth century, therefore, there were two main ways of trying to apply homeopathy. One was to keep reading descriptions of the effects of medicines and wait until you found a patient suffering from the corresponding symptoms, and the other was to take the symptoms of your patient and look them up in a repertory to see which medicines seemed to suit them. In practice homeopaths used both methods.

Obviously all this depended critically on the reliability of the materia medica. Some homeopaths, especially Richard Hughes in England, became unhappy on this score, for two main reasons. First, homeopathic authors had an unfortunate tendency to copy from one another uncritically. Second, the literature was coming to be include more and more "clinical symptoms" — symptoms that were not based on provings. If a patient recovered after receiving a homeopathic medicine the prescriber might record the fact in print and the patient's symptoms could then become attached to the medicine in question even though they had not appeared in provings. While this might well be useful in practice it represented a watering-down of the original homeopathic idea. At first these clinical symptoms were distinguished in the reference books by a special mark, but soon this was omitted and the homeopathic literature moved a further step away from its early base in the provings.

As we shall see, Hughes made a gallant but ultimately unsuccessful attempt to purge the materia medica of what he regarded as unreliable information and to bring it back to the proper path of provings and toxicology.

Potency

Even in Hahnemann's lifetime homeopaths were divided on the potency question and this division persisted after his death. Indeed as scientific knowledge advanced the problem became more acute, for it grew increasingly difficult to think of plausible explanations for the supposed activity of very dilute solutions. Hahnemann had not recognised this difficulty. He reasoned that however much a substance might be diluted there must logically be some of it still there, and this should be enough to produce an effect. But

Hahnemann lived just before the development of modern molecular theory.

The chief architect of this theory, Avogadro, had in fact published his theory in Hahnemann's lifetime but it is unlikely that it came to his attention. According to the modern understanding based on Avogadro's work, matter is not infinitely divisible as Hahnemann supposed. If a substance is diluted progressively in the Hahnemannian manner there must come a time when the solution no longer contains any molecules of the original substance at all. Theoretically this should occur at about the 24th decimal or 12th centesimal potencies, which are therefore regarded by modern homeopaths as a watershed between low and high potencies. So-called ultramolecular potencies are those above this level.

Scientifically minded homeopaths in the late nineteenth century were deeply troubled by the potency question and made experiments to try to find out what happened when substances were triturated or diluted. They discovered that when metals were triturated they could be detected under the microscope up to the 12th decimal in some cases although only up to the fourth or fifth in others. They concluded, not unreasonably, that the finely divided particles should have an enhanced effect inside the body owing to the relative increase in their surface area. In this view they were supported by a most eminent physicist, Professor Doppler of Prague, though he did not refer explicitly to homeopathy.

Other scientific facts were adduced in support of the idea that small doses could have an effect on organisms. Frog's semen had been shown to be capable of fertilising frog's eggs when diluted to one part in a million, and a one in a hundred dilution of cowpox serum produced infection in children vaccinated with it. These and similar reports encouraged homeopaths but many of them rejected Hahnemann's contention that so-called dynamisation actually increased the power of the medicines, and one, named Veith, explicitly recognised the metaphysical nature of the theory, saying that it was a new application of one of the doctrines of the founder of the Iranian religion, Zoroaster.

While scientifically minded homeopaths were trying to investigate the potency idea by the accepted methods of science, the more extreme homeopaths had adopted Hahnemann's teachings uncritically and enthusiastically and had indeed gone much further than had the Master himself. However eccentric Hahnemann may have become as he aged he continued to preserve a streak of caution and common sense.

Although he had laid down the rule that the standard potency for all purposes — treatment and provings — was to be the 30th centesimal, he went on using a variety of potencies including on occasion much lower ones, and at his death his medicine case was found to contain at least one bottle of an undiluted tincture. He experimented with the 60th and even the 300th potencies, but no higher, and when he was told that one of his disciples, von Korsakoff, had gone much higher — up to the 1500th — he contented himself with remarking that the only importance of this was to show how far it was possible to take potentisation without loss of effect. Nevertheless, as he sagely remarked, "there must be some limit to the thing".

Some of his followers refused to recognise any limits and went far beyond even von Korsakoff's 1500th potency. Of course, to produce even a 1500th potency by hand takes a long time, but a way round this difficulty was discovered by another ingenious homeopath, Julius Caspar Jenichen, one of the most colourful of the early recruits. Like a number of homeopaths of the day he was not a doctor; he had been horse-trainer (or Master of Horse, depending on whom you believe) to the Duke of Gotha.

He was a man of enormous physical strength, which he used to display at dinner parties by rolling up silver salvers and tearing them in half; a habit which, as a contemporary remarked, somewhat diminished his appeal as a dinner guest. On taking up homeopathy he applied his physical prowess to the manufacture of homeopathic medicines, shaking the vials so hard that they "rang like a bell." During his lifetime he kept his methods secret, but from notes left at his death (by suicide) it appears that he based his practice on an idea that Hahnemann had held at one time but later abandoned: namely that what matters is not the dilution of the medicines but the number of times they are shaken.

Jenichen arbitrarily decided that ten shakes were equivalent to one degree of potency, and starting from the 29th or even lower degrees he went on to make much higher potencies than anyone else had done. There is a suggestion that he diluted the medicines after every 250 shakes but this is uncertain. In any case, what mattered, it seems, was the force that Jenichen was able to apply. By the time of his death he had obtained potencies as high as 60,000.

Jenichen's methods were adopted enthusiastically by certain homeopaths, including von Boenninghausen in Europe and Constantine Hering in America. Hering lamented that it would be hard to find anyone after Jenichen who would be physically capable of preparing such high potencies, but happily Jenichen had left a supply of medicines large enough to serve the next two generations of homeopaths. In any case Hering's fears proved unfounded, for shortly afterwards American homeopaths applied New World know-how to the problem and invented various kinds of potentising machines that allegedly took potency to much dizzier heights even than those scaled by the intrepid Jenichen.

The advocates of ultra-high potencies did not make any attempt to verify their claims scientifically. Their own practical experience, they felt, was enough. They did, it is true, make a perfunctory bow in the direction of science: Hering suggested that potency was due to a new and hitherto unheard-of natural force, which he called Hahnemannism on the analogy of Mesmerism and Galvanism. Glass and cork, he alleged, were insulators of Hahnemannism as they are of electricity. Scientifically minded homeopaths, however, treated these ideas with ridicule.

In later years what might be called the scientific and the spiritual wings of homeopathy, were to become identified with low-potency and high-potency prescribing respectively, though the differences between the schools went much beyond this and affected their whole approach to homeopathy.

The impact of orthodox medicine

By the end of the nineteenth century the old school of medicine that Hahnemann had fought so implacably was itself on the decline — not primarily because of homeopathy but owing to advances in scientific knowledge. Pasteur and Koch had proved that some diseases, at least, were caused by microbes, while Virchow claimed that disease could be understood by considering the body as a commonwealth of cells — an idea that contained the seed of the ultimate destruction of vitalism as a scientific concept. Chemists too were helping in the understanding of the way the body functioned in health and disease. Altogether it was a most exciting time intellectually; doctors felt that real advances in medicine were being made for the first time in centuries.

These developments posed a serious problem for homeopaths, in view of Hahnemann's rejection of the possibility of understanding the mechanism of disease. Should homeopathy stand fast on this or should it move with the times? Some homeopaths held rigidly to Hahnemann's teaching and rejected the new knowledge as untrue or irrelevant, while others yielded to its seductions and tried to reinterpret homeopathy in its light.

It was especially on the continent of Europe that attempts were made to reconcile homeopathy with the new medicine. Some of these attempts were linked to contemporary notions of biochemistry. A German homeopath, von Grauvogl, believed that people could be classified in three constitutional types, according to whether they had an excess of water, oxygen, or carbon and nitrogen in their tissues. Various derivatives of this theory are still influential in French homeopathy today. (Note that this theory of constitution is quite different from that discussed in Chapter 9, which has other origins.)

Another nineteenth-century idea still active in French homeopathy today derives from Rademacher, a contemporary of Hahnemann. He taught that disease results from disordered functioning of various key organs, such as the liver and kidneys, and that medicines should be given to "drain" them. Although Rademacher was not a homeopath his ideas were adopted by some homeopaths as a basis for treating chronic disease. In France homeopathic medicines in low potency were — and still are — given as "drainage remedies".

Yet another approach to prescribing was suggested by Schussler. He postulated that the cause of disease is disturbance in the concentration of various salts within the body cells, and he held that these disturbances could be corrected by means of his twelve "tissue salts", which are low-potency preparations of various inorganic compounds. They are still available today.

It need hardly be said that the theoretical foundations of all these nineteenth-century systems have long been rendered antiquated by later discoveries. They survive partly because — for whatever reasons — they appear to work for some people, and partly because they help to simplify the complex business of choosing homeopathic medicines in chronic disease. They have however never been accepted by homeopathic purists, and they are best regarded as offshoots from the main homeopathic trunk.

Chapter 6

Early English Homeopathy

In 1826 a young English physician, Frederic Hervey Foster Quin, visited Leipzig. He had been travelling for some time on the Continent for the sake of his health and his journey to Leipzig was prompted by an interest in homeopathy that had been planted in his mind by a fellow physician, Dr Neckar, who had treated him.

Quin seems to have been very well connected; in fact he is rumoured to have been the natural son of Elizabeth ("Bess") Cavendish, who became Duchess of Devonshire when she married the Duke after the death of his first wife, Georgiana. This theory has been disputed but is supported by the fact that her maiden name was Hervey and her first husband was called Foster. Possibly thanks to the Duchess's influence Quin was appointed personal physician to Napoleon during his exile on St Helena, but his patient died before he could take up his duties.

By 1826 Hahnemann was living in Köthen. It seems that Quin visited him there, but it was mainly the example of the Leipzig homeopaths that impressed him. By now largely though not wholly converted to the new system he went to practise it in Paris. He spoke French fluently and was an enthusiastic francophile. In 1832 he returned to London a completely dedicated homeopath.

Quin was soon very successful in London, in spite of the hostility of the Royal College of Physicians, a number of whose members blackballed him for membership of the Athenaeum. Thanks to his aristocratic connections, however, homeopathy prospered, and indeed it was owing to Quin that the new system first attracted royal patronage. The Prince and Princess of Wales were among those who visited him on his deathbed.

In 1844 Quin founded the British Homeopathic Society and in 1850 a homeopathic hospital was opened in Golden Square, Soho. This was the forerunner of the present hospital in Great Ormond Street, which was bought by subscriptions from wealthy patrons. Lord Grosvenor, later Lord Ebury, became the first chairman of the hospital board. A bazaar for the hospital was held in 1857 at the Riding School of the Cavalry Barracks in Hyde Park; the items donated for sale included a contribution from the famous artist Sir Edward Landseer.

In 1854 London was struck by an outbreak of cholera. This gave homeopaths a chance to show what they could do. Among the patients admitted to the orthodox hospitals the death rate was 52 per cent, while at the homeopathic hospital, where 61 patients were admitted, only 10 died (16 per cent) — and of these, one died at the door of the hospital as he was being taken from the cab and another was treated only after he had been given up by an orthodox physician. The Board of Health and the Medical Council omitted the figures for the homeopathic hos-

Figure 6.1: Dr Frederic Quin

pital in the Blue Book published in 1855 to report on the outbreak, but Lord Grosvenor raised the matter in the House of Lords and a report including the homeopathic results was subsequently published.

At this distance in time it is impossible to know why the cholera patients treated homeopathically did better. (Similar results were reported at this time for Italy.) Today we know that the essential treatment for cholera is fluid replacement, and if this is done the death rate is low; but neither homeopaths nor conventional doctors would have been aware of this in the nineteenth century. Perhaps the better results achieved by the homeopaths were due to their patients' not receiving conventional treatment.

The hospital was supposed to be purely homeopathic, and at a meeting in 1870, with Quin in the chair, it was decreed that medicines other than those listed in the homeopathic pharmacopoeias were not to be kept, and even the prescribing of undiluted tinctures was discouraged.

Nineteenth-century English homeopathy

What may be called the English school of homeopathy in the nineteenth century produced two writers of outstanding importance, Robert Dudgeon and Richard Hughes.

Dudgeon was an early recruit to the homeopathic banner raised by Quin. A German scholar, he translated nearly all Hahnemann's writings into English and kept closely in touch with what German authors of the day were saying on the subject. Thanks to him we have a good insight into homeopathy in the immediate post-Hahnemannian era. Though a convinced homeopath himself, Dudgeon was not afraid to voice his own opinion or to criticise the Master where he felt it to be appropriate. He had a scientific bent and invented a machine for recording changes in blood pressure. He had a pleasantly ironic sense of humour and is one of the most stimulating and readable of the early homeopathic writers.

Important though Dudgeon's contribution was, however, it was his friend and colleague Richard Hughes who did most to shape British homeopathy at this period. Although he was at one time on the staff of the London Homeopathic Hospital, Hughes spent most of his medical career in practice in Brighton, though it is difficult to believe that he had a lot of time to spare for actually seeing patients.

He organised the five-yearly International Homeopathic Congresses and he edited the *Annals* of the British Homeopathic Society. His most important and influential role, however, was as a teacher and writer. He was appointed Lecturer in Materia Medica by the British Homeopathic Society and his lectures were published and used as the basis for instruction of doctors up to his death in 1902. His views on homeopathy were endorsed by Dudgeon and others as an authentic up-to-date interpretation of homeopathy. Hughes became in fact the Grand Old Man of British homeopathy in the nineteenth century (though to be sure he was only 62 when he died). It is therefore legitimate to speak of Hughesian homeopathy, though it must be understood that this was not Hughes's view alone but was the orthodox British homeopathy of the day.

Hughesian homeopathy

The essential character of Hughesian homeopathy was that it lay at the "scientific" end of the homeopathic spectrum of opinion. That is, it was pragmatic and anti-mystical. On the theoretical level Hughes, Dudgeon and other leading British homeopaths of the day rejected Hahnemann's concept of the vital force, his theorising about how homeopathic medicines worked, and the psora theory. They were also unhappy about potency. They were prepared to concede that, in practice, some high dilutions — at least up to the 30th centesimal — did seem to work, but they recognied the difficulty of explaining this in terms of the contemporary knowledge of physics and chemistry. The vast majority of British homeopathic prescribing at this time was based on the use of very low (material) dilutions — 6c and below. As for the claims of Jenichen, Hering and others to be able to produce ultra-high potencies by various non-Hahnemannian techniques, Hughes and Dudgeon treated these with gentle derision.

As a homeopath Hughes naturally placed the similia principle at centre stage but his attitude to it was relaxed and non-dogmatic. It was, he said, not a law of nature as Hahnemann claimed but simply a rule of thumb — a skeleton key to try in the therapeutic lock. It often gave the right answer but not invariably, nor was it the only key worth trying. For some disorders, such as angina, Hughes thought that allopathy was more effective.

Hughes believed, moreover, that if you are serious about the similia idea you must take pathology into account. It was all very well for Hahnemann to say that nothing could be known about the causes of disease; in his day that might have been true, but times had changed and quite a lot was now known about pathology and the new knowledge needed to be incorporated into homeopathy. Hughes believed that medicines should be chosen not just on subjective symptoms they produced but on the basis of their known pathological effects on human beings and even (daringly) on animals. For example, if your patient is suffering from an ulcer you should choose a medicine known to produce ulcers, and so on. This insistence on the role of pathology in prescribing was to cause later generations of homeopaths, who were following a very different star, to adopt a superior attitude to Hughes and to label him pejoratively as a mere "pathological prescriber".

Hughes also insisted that, in trying to find the correct remedy you should consider the time sequence in which symptoms occur. It was not enough to look simply at lists of symptoms recorded in the provings. Each disease had its characteristic sequence of symptoms and you should try to find a remedy with a similar pattern of symptom development. Unfortunately, owing to the way Hahnemann had recorded the results of his provings, you could not do this from his published work.

Important though all these ideas were for British homeopathy, what really distinguished Hughes was his critical and scholarly approach. Most homeopaths of the day outside Britain, especially in America, based themselves on Hahnemann's later work almost exclusively — that is, on the fifth edition of *The Organon* and on *The Chronic Diseases*. Hughes, in contrast, looked at Hahnemann's writings as a whole. He carefully charted the way the Master's thought had evolved over the years and was not afraid to say in what ways he thought it had changed for the worse. He pointed out, for example, that Hahnemann's laying down the rule that the 30th potency should be used for all purposes had fossilised homeopathy most undesirably. He also showed that the so-called provings of *The Chronic Diseases* could not have been carried out in the same way as those of *The Materia Medica Pura* and so could not be relied on as accurate descriptions of the effects of the new medicines. Such views, of course, were lèse-majesté in the view of the large number of homeopaths for whom Hahnemann's words were law.

Hughes's contribution to homeopathy was not confined to critical discussion of Hahnemann's writings. His most important undertaking was undoubtedly his attempt to revise and purify the homeopathic materia medica, which resulted in his writing the rather ponderously titled *Cyclopaedia of Drug Pathogenesy*. Hughes had earlier collaborated with the American T.F. Allen in the production of that editor's *Encyclopaedia*, but later he came to feel that Allen had been too uncritical and had included much that would have been better omitted. Hughes's aim was to sift all this material and publish only what he thought was reliably established.

This was a truly monumental undertaking. The four volumes of the *Cyclopaedia* took seven years to prepare (1884-91). It was a joint enterprise, in which the British Homeopathic Society collaborated with the American Institute of Homeopathy; nevertheless the impetus behind it came from Hughes and he carried out most of the work. His intention was to include all the reliable information available in his day apart from that in Hahnemann's writings. This involved a vast amount of translating, sifting and editing.

A number of rules were adopted to eliminate untrustworthy reports. No purely "clinical" symptoms were included, of course, and nor were symptoms obtained with high dilutions (above 6c) unless confirmed by provings of more material doses. A very important feature was that all the provings were given in narrative form so that the way in which the symptoms had developed could be read and understood.

The *Cyclopaedia* was a unique attempt to present a truly critical collection of the materia medica and demanded a high degree of dedication from its readers. Even though the symptoms were presented in narrative form rather than as lists, they were so compressed that they were hard to take in. Hughes was evidently sensitive on this score, for he wrote: "It seems to be the impression of some that our *Cyclopaedia* is a mere luxury of pathogenesy, quite beyond the requirements of the student and the practitioner, and only really valuable to the teacher or writer on the subject." But it was the student who was expected to use the *Cyclopaedia*. Thanks to it the subject "will be found full of life and meaning; and materia medica, hitherto the dullest and most hopeless, will become the most interesting of studies".

Hughes's contemporaries shared his enthusiasm. At his death an obituarist in the American *Hahnemannian Monthly* described the *Cyclopaedia* as "a work without parallel in all medical literature" (which was undoubtedly true) and went on to say: "It is a work — we had almost said THE work — from which the future materia medical authority will compile all that is best and most reliable in his new textbook; and it requires no prophetic vision to foretell that its pages will be even more frequently explored at the end of the twentieth century than at its beginning."

Alas for prophecy. Within a few years of Hughes's death his *Cyclopaedia*, together with the rest of his work, had been forgotten almost as if it had never been, and later generations of homeopaths were to drink from a very different source.

To some extent this surprising turn of events can be explained as a natural reaction by British homeopaths against the ideas of a man whose influence had been paramount for so many years. Hughes was in many ways open-minded and undogmatic but it was no doubt inevitable that his teaching would eventually harden into a kind of orthodoxy. Paradoxically however it was Hughes's very absence of dogmatism that made him seem to some later homeopaths a traitor to the cause, for it led him to minimise the differences that separated homeopathy from orthodox medicine.

It took considerable courage for a doctor to declare himself a homeopath in Hughes's day; nevertheless Hughes seems to have felt no reciprocal hostility for his orthodox opposite numbers and indeed, in his last published work, *The Principles and Practice of Homeopathy*, he made a remarkable plea for reconciliation. He was well aware, he wrote, of the many shortcomings of homeopathy and of the "fancies and follies" that had become incorporated in it. What was needed, he said, was for orthodox doctors to bring their resources of time, expertise, and intellect to bear on homeopathy and help to put it on a sound scientific footing.

Hughes himself had no doubt about where such a change would lead.

> Do our brethren know what would be the result of such generous policy? We should at once cease to exist as a separate body. Our name would remain only as a technical term to designate our doctrine; while "homeo-pathic" journals, societies, hospitals, dispensaries, pharmacopoeias, directories, *under such title*, would lose their raison d'etre and cease to exist. The rivalry between "homeopathic" and "allopathic" practitioners would no longer embitter doctors and perplex patients.

I suspect that it was this wish to unite homeopathy with orthodoxy, rather than his more technical views about the right way to choose medicines, that was the real reason for the virtual suppression of Hughes's ideas by later homeopaths. If Hughes had succeeded in effecting a reconciliation between homeopathy and orthodoxy it is likely that — as Hughes himself realiaed — the result would have been the disappearance of homeopathy as a separate form of medicine; this did in fact happen later in the USA.

To a modern doctor Hughes's writings and those of his friend Dudgeon are among the most accessible of homeopathic texts, not excepting those of the twentieth century. Although the medical ideas with which these authors worked are long out of date, their pragmatic and critical attitude makes them surprisingly modern in tone and readable even today. Nevertheless after Hughes's death British homeopathy moved decisively away from science, and Hughes himself received the contemptuous Hahnemannian label of "half-homeopath". In subsequent chapters I shall look at the reasons for these developments.

Chapter 7

Homeopathy in America

The story of the rise, decline, and fall of homeopathy in the USA is a fascinating subject in its own right, but it has a significance that is of more than purely American importance, for it exemplifies both the strengths and the weaknesses of homeopathy as a medical doctrine. Moreover the story is very important for the subsequent development of homeopathy, for its sojourn in America changed it profoundly in several ways and these changes were later exported to other countries, notably Great Britain.

The first homeopathic doctor in America was Hans Gram, an American of Danish extraction who settled in New York in about 1825 and converted a number of other doctors to the new system. Much the most important homeopath of the period, however, was Constantine Hering, whose labours established homeopathy as an important feature in the American medical scene.

Hering was a German, born in 1800, who as a medical student was entrusted by his tutor with the task of writing an attack on homeopathy. But his researches led to his conversion. Soon after qualifying in 1826 he joined an expedition to South America, where he practised homeopathy and conducted provings, including some interesting ones on snake venom. In 1833 he returned briefly to Germany, but on his way back to South America he called at Philadelphia, where he was persuaded to stay on. He remained in the USA, except for one year in 1845, for the rest of his life, and became the Grand Old Man of American homeopathy, practising the discipline, conducting provings on himself and others, writing, and organising. He was elected first president of the American Institute of Homeopathy when it was founded in 1844.

Throughout the nineteenth century homeopathy prospered strongly in its new home. Homeopathic colleges sprang up all over the country, and many thousands of practitioners graduated through them. The most famous of these establishments was the Homeopathic Medical College of Philadelphia, but there were also many others; in 1900 there were 22 colleges, and before the First World War there were 56 purely homeopathic general hospitals, some with up to 1400 beds. There were 13 mental asylums with up to 2000 beds each, 9 children's hospitals, and 21 sanatoriums. This degree of public acceptance of homeopathy could be matched nowhere else in the world.

Figure 7.1: Constantine Hering

The reasons for the early success of homeopathy in the USA are not difficult to understand in the context of the state of orthodox medicine at that time. Mainstream American medicine in the first half of the nineteenth century was comparable with European medicine but was if anything more dangerous. Bleeding was of course a sovereign remedy and was taken to even greater lengths than were fashionable in Europe. A prominent physician of the time, a Dr Rush, wrote that blood should if necessary be let until four-fifths of the blood in the body had been removed. Whereas the English physician Sydenham recommended that a mere 40 ounces of blood be removed in the treatment of pleurisy, the redoubtable Rush held that for the more virulent form of pleurisy encountered in the USA at least double that amount must be withdrawn. The death of George Washington provides an appalling example of this. The unfortunate President suffered from a sore throat and was bled repeatedly by his attendants until no more blood could be extracted, when he died.

Blood-letting was the correct treatment for almost any disease but especially for fever. A textbook of 1836 devoted 33 pages to various techniques of phlebotomy and another, published in 1847, had no less than 87 pages on this all-important subject. Children were supposed to stand in particular need of this form of treatment and here once again the egregious Dr Rush expressed his opinion forcibly. In answer to the charge that blood-letting was killing many children, Rush replied that he could mention "many more instances in which blood-letting has snatched from the grave children under three or four months [sic] old by being used from three to five times in the ordinary course of their acute diseases". A professor at the College of Physicians and Surgeons in New York, writing in 1840, upheld the importance of bleeding the very youngest children and indeed the newborn, though he did remark regretfully that "the young subject does not bear the loss of considerable quantities so well as the adult" and that there was an increased tendency for convulsions to occur.

Another form of treatment that almost rivalled blood-letting in popularity was dosing with calomel (mercurous chloride). As might be expected, this was firmly advocated by Rush in almost every disease. Calomel had been used for many years but mainly in the treatment of chronic diseases including syphilis. At the end of the eighteenth century Rush introduced it to treat an epidemic of yellow fever in Philadelphia. The rationale for its use was chiefly that it acted as a purgative and so would rid the body of the toxic substances that caused the disease. Naturally Rush advocated enormous doses. Equally naturally the patients suffered serious, sometimes fatal, mercury poisoning. There are horrific accounts of patients who lost eyes and ears and large quantities of flesh from their faces before they died; others, slightly more fortunate, survived but often at the cost of most of their teeth and part of their jaws, for they frequently suffered contractures of the jaws that necessitated extensive surgery to allow them to eat.

For understandable reasons, many patients in the early part of the nineteenth century were turning away from orthodox medicine to various forms of folk medicine, especially indigenous herbalism.

The unorthodox systems, however, were the province of unqualified practitioners, many of whom were barely literate. Homeopathy, the new arrival from Europe, had the advantage of being practised by qualified doctors who were in many cases better educated than their orthodox rivals, for at this time the homeopathic literature was almost all in German or Latin and so could only be read by men with a mastery of those tongues. The presence of a large number of German immigrants to the USA at this time also helped in the spread of the new system.

The homeopaths' success naturally excited the hostility of the orthodox physicians and numerous criticisms of homeopathic theory and practice appeared. The foundation of the American Medical Association was at least in part a reaction to the success of homeopathy; doctors professing themselves to be homeopaths were not admitted to membership. On at least one occasion homeopathic and orthodox physicians attempted to settle their differences by resorting to fisticuffs. But homeopathy continued to prosper, reaching its peak of success after the Civil War, in the decades 1865-85. Gradually orthodox hostility lessened; by the mid-1880s homeopathy had largely ceased to be on the defensive and its future seemed assured. In reality, however, it was about to go into decline.

Decline and fall

There were two main reasons for the decline of homeopathy in America after about 1885. One was, paradoxically, the very success of the new system. Orthodox physicians were coming to realise the dangers of existing methods of treatment and were beginning partly to abandon the use of large doses of drugs and of bleeding. The example of homeopathy undoubtedly played a part in prompting this trend; nor was it only in the matter of dosage that orthodox medicine borrowed from homeopathy. Some of the homeopaths' drugs began to find their way into the orthodox pharmacopoeia while others had always been common to both schools, and this tended to blur the distinction between them still further.

The second reason for the decline of homeopathy was the old problem of dissent within the ranks of the faithful. As the hostility of the orthodox school decreased many homeopaths made various compromises with orthodoxy, using conventional drugs at times, giving material doses, taking account of pathology, and ignoring the doctrines of psora and vitalism. Against these "half-homeopaths" a small but resolute body of purists held out for the extreme position adopted by Hahnemann in his later years. As nearly always happens within a heretical sect, the virulence with which the two factions attacked each other far exceeded their hostility towards their orthodox opponents.

The banners under which the factions assembled were, respectively, those of high and low potency prescribing, but the grounds for disagreement between them were much wider than this and extended to almost every aspect of homeopathy. Matters came to a head in 1870 at a meeting of the Institute, at which Carroll Dunham, the president, made an important speech. To be a homeopath, he said, required adherence to a fundamental therapeutic law, but there could be disagreement about its detailed interpretation. He himself was a purist, a rigid Hahnemannian; nevertheless he had to acknowledge the existence of self-styled homeopaths who thought otherwise, and the right way to deal with them, he believed, was not to proscribe them but to encourage free and open discussion.

Dunham's tolerance was admirable but its effects were the reverse of what he intended. The argument, which had hitherto smouldered underground, now burst out in the open and became much fiercer. Sporadic attempts were made to establish a set of articles to which all would-be homeopaths must subscribe but this was not accepted. The Institute grew rapidly in numbers but the new members lacked the proselytising fervour of the old guard, whom they looked on as obscurantist old German fuddy-duddies. The purists, for their part, regarded the new recruits as upstarts who were ignorant of materia medica, did not know how to individualise their cases, had never read *The Organon*, and did not even believe in the law of similars. Low-potency and high-potency journals appeared to cater for the two camps. Rival homeopathic societies and even rival homeopathic hospitals appeared, and the public naturally found the situation puzzling and unsatisfactory.

The low-potency group, which had always greatly outnumbered its rivals, drew gradually closer and closer to orthodoxy. Eventually the distinction between homeopathy and allopathy became so slight that there seemed no point in perpetuating it, and the vast majority of American homeopaths quietly switched their allegiance. By 1918 the number of homeopathic colleges had declined to seven, and before long these too had disappeared. The Homeopathic Medical College of Philadelphia stopped teaching homeopathy in the 1930s, by which time homeopathy had ceased to be a live issue in American medical politics; it was in fact as good as extinct.

What I have just described is an outline of what might be called the "political" rise, decline and fall of homeopathy in America. The story has however another dimension, which is of the greatest importance for the development of homeopathy down to the present day. In America homeopathy became fused with Swedenborgianism to give a hybrid growth that differs in several important ways from Hahnemannian homeopathy but is today widely taken to be the original doctrine. But important though it is, most people who have written about homeopathy have either ignored the Swedenborgian element or have played it down. I shall therefore fill in some of the gaps, but before doing so I need to digress briefly to give an outline of Swedenborgianism, since this is likely to be unfamiliar to most readers.

Swedenborgianism and homeopathy

Emanuel Swedenborg (1688-1772) is a most extraordinary figure. A scientist, engineer, statesman, and philosopher, who achieved great distinction in his own country, Sweden, and renown abroad, he showed remarkable wisdom in the practical management both of his own affairs and those of his country. And yet from middle age onwards he had what he believed were continual contacts with the spirit world, for the most part not in trance but in full consciousness. From these experiences he was able to construct a complete cosmography of the spirit world and its relation to our own. Nor was this all, for in 1743, in Amsterdam, he had a profound mystical experience that became the starting point for a thorough re-evaluation of the whole of religion and eventually led him to undertake a detailed allegorical interpretation of much of the Old Testament.

Yet his career began conventionally enough. In 1724, when he was 36, Swedenborg became an assessor for the Board of Mines.

The work took him all over Sweden and led him to make numerous important scientific studies of mineralogy and other matters. He showed remarkable technological ability but this was complemented by a wide philosophical outlook. In 1736 he obtained leave of absence to go abroad; he went to Paris to study anatomy, not intending to become a doctor but hoping to gain insight into the relation between mind and body. This experience proved a decisive turning-point in his life and resulted in the publication of his profound and far-reaching book, misleadingly entitled in English *The Economy of the Animal Kingdom*, which is really a synthesis of scientific and mystical views of man and the world.

After completing his anatomical studies Swedenborg returned to his duties at the Board of Mines in Stockholm for a time, but in 1743 he was abroad again in Holland and England. From 1745 onwards his conversations with spirits began in earnest, and he came to believe that he had received a divine commission to reinterpret the Bible. Henceforth he led a double life: outwardly he continued to be the practical man of affairs, while inwardly he went on exploring the spirit world. He was able to give concrete evidence of the reality of these experiences: on one occasion he apparently had clairvoyant knowledge of a fire in Stockholm when he was 300 miles away in Gothenburg. This so impressed the philosopher Kant that he took pains to investigate the authenticity of the story, and was wholly convinced by the result of his researches. Other apparently well-authenticated instances of Swedenborg's paranormal abilities concern a secret known only to the Queen of Sweden, which was revealed to Swedenborg by a spirit, and telepathic awareness of the death of Czar Peter III in prison.

In 1747 Swedenborg finally resigned from the Board of Mines to devote himself entirely to his major work on the Bible. This was published, anonymously at first, in London in eight large volumes from 1749 to 1756, under the title *Arcana Coelestia* ("Heavenly Secrets"), and was almost entirely ignored. In 1751 Swedenborg was back in Sweden, indulging in his long-standing passion for gardening. He also wrote practical and very sensible memoranda on economic problems such as inflation for the Swedish Diet, of which he was a member. He also continued to travel and to write on religious and mystical matters. He died peacefully at the age of 84 in London;

an appropriate resting place, for England was always his favourite country.

This is not the place to attempt an assessment of a man of such richness and complexity of character as Swedenborg, but it's worth remarking on the charm, intelligence, and lack of fanaticism that come through from his life and writings and on the eminent good sense that he showed in practical matters of every kind. The psychiatrist Henry Maudsley wrote a paper on Swedenborg describing him as schizophrenic, but there is no sign of this except perhaps in respect of his mystical ideas; and, as the philosopher C.D. Broad remarked, if these were delusions they were at least grafted on a mind that in every other way was remarkably sane.

During his lifetime Swedenborg came under attack from orthodox churchmen — by no means a trivial matter at that time, when to be tried for heresy was still a real risk — but he successfully withstood these threats. Although he believed that he had been the vehicle for a new religious revelation, he did not found an organisation to carry on his teachings. After his death, however, a New Church dedicated to the preaching of his ideas was founded in England, where Swedenborg's ideas had an influence on Blake and Coleridge among others. The New Church quickly fragmented into at least three groups, and this tendency to schism continued to characterise it when it crossed the Atlantic to America, which it quickly did. Nevertheless it was successful in the USA, and soon after its introduction in 1784 it established itself in a number of American cities.

The appeal of Swedenborg's ideas is not hard to understand. By the beginning of the nineteenth century the rapid advance of science was coming to be seen as a threat to established religion. Darwinism still lay in the future, but the intellectual climate that was to provide the environment for the fateful clash between science and religion already existed. People were asking questions and were increasingly unwilling to be told that the answer lay in faith.

Swedenborg, too, rejected faith, at least in the ordinary sense of the word. He claimed that his teachings were based on direct revelation but he by no means despised reason. He was a perhaps unique combination of mystic and scientist and his ideas were particularly attractive to intellectuals who wished to preserve a religious attitude yet were aware that advances in scientific knowledge were radically altering ways in which people thought about the world. The novelist Henry James and his brother William, the philosopher and psychologist, were brought up as Swedenborgians.

Swedenborgianism and homeopathy took to each other at once. Swedenborgians found in homeopathy a medical system that perfectly complemented their religious attitude, while homeopaths found in Swedenborgianism a religious framework into which Hahnemann's ideas could expand freely. Homeopathy thus quickly became the accepted medical system for Swedenborgians, while most of the leading nineteenth-century homeopaths, including Hans Gram and Constantine Hering, were Swedenborgians. The firm that came to dominate the homeopathic drug industry after 1870 was that of Boericke and Tafel of Philadelphia, whose owners were Swedenborgians; the same men also became the leading homeopathic (and Swedenborgian) publishers in the USA.

The features of homeopathy that made it so congenial to the Swedenborgians were the very ones that disturbed "scientific" homeopaths in England like Dudgeon and Hughes, for it was naturally the ideas of Hahnemann's late phase that appealed most strongly to the Swedenborgians — vitalism, the miasm theory, potentisation, and the divine inspiration of the similia principle. All these ideas were adopted by the Swedenborgians and taken to new lengths.

For Swedenborg the idea that there is a mystical correspondence between the spirit world and our own was fundamental. Like many earlier thinkers, including the alchemists, Swedenborg taught that the form and function of man (the microcosm) is modelled on, and reflects, that of heaven (the macrocosm). The alchemists, taking their cue from the divine Egyptian originator of their craft, Hermes Trismegistus, were wont to repeat the phrase "as above, so below". Swedenborg likewise held that whatever happens in the spirit world must have its counterpart here on earth. This idea of correspondence could easily be linked with the similia idea, and it was natural for the Swedenborgians to regard this as a divinely ordained law of nature.

Vitalism, likewise, was wholly congenial to the Swedenborgians. Swedenborg held that the essential nature of a man is determined by his "will" and "love" — that is, by his basic spiritual impulses. This teaching could be directly equated with the Hahnemannian notion that disease is caused by derangement in the vital force. But the Swedenborgians took the idea further than Hahnemann had done, maintaining that disease always begins at the inmost spiritual level — that of the will and understanding, around which the physical body accumulates rather as the caddis worm builds its house of stones or bits of wood. Disease is the reflection of a failure on the part of the builder — it results from a disorder of the will or understanding and is thus a moral as well as a physical problem. It follows that the homeopath must not treat the patient's body alone but also his mind and inner spiritual essence.

The Swedenborgian homeopaths gave a definite moral twist to the miasm theory. For Hahnemann the miasms had been acquired "infections", but for the Swedenborgians they were moral taints passed from generation to generation, and psora in particular took on the characteristics of Original Sin. At the same time, however, and somewhat inconsistently, the miasms continued to be thought of as somehow invading the organism from without and progressing inwards until they finally reached the soul.

The American translator of *The Chronic Diseases,* Charles J. Hempel, was quite explicit about the connection with Original Sin and even gave homeopathy a millenarian twist. Psora will eventually be completely eliminated from mankind thanks to homeopathy, he claimed, and the consequent "complete physical regeneration of human nature" will bring about a total transformation of society! Radical ideas of this kind have continued to appeal to some homeopaths down to our own day; I touch on this again at the end of Chapter 12.

In 1865 Hering wrote a very influential article based on the psora doctrine. He claimed that as a disease becomes chronic the symptoms always move in a particular way: from the surface to the interior, from the extremities to the upper part of the body, and from less vital to more vital organs. On the basis of this alleged progression of symptoms he propounded his "Laws of Cure", which state that cure must take place in the reverse order to the march of the symptoms: that is, from within outwards, from above downwards, from most important to least important organs, and in the reverse order of their appearance (first in, last out). The development of a rash in treatment, for example, is a favourable sign (because "the psora is coming out"), and the same applies to the reappearance of symptoms from which the patient has not suffered for many years.

Hering's laws were arrived at largely on theoretical, *a priori* grounds, but they were quickly incorporated into homeopathic doctrine in America and numerous confirmations of them were reported — which was hardly surprising, since Hering had set up a "heads I win, tails you lose" method of confirming the theory. Any cure that failed to follow the prescribed sequence was automatically discounted as mere palliation while every case that obeyed the laws was quoted as proof of their truth.

Hering also developed the miasm theory in another way, by recognising the existence of other miasms in addition to sycosis, syphilis, and psora. Almost any disease could be looked on as a potential miasm — that is, as capable of leaving long-lasting taints in people who had once suffered from it. The products of that disease could then be used "isopathically" to treat the patient.

The idea of isopathy is to take the thing that causes the disease and potentise that to use as a medicine. So potentised grass pollen is isopathic, because it causes hay fever in susceptible people. Allium cepa, derived from onions, is also used to treat hay fever, but it is homeopathic rather than isopathic because onions do not cause hay fever.

Another idea which developed at this time was the use of pathological material as remedies. The American homeopaths took substances such as gonorrhoeal pus or tuberculous lung tissue, potentised them in the Hahnemannian manner, and used the resulting medicines to treat patients who had symptoms suggestive of the diseases in question. So we get Medorrhinum (from gonorrhoea) and Tuberculinum, for example. Some of these nosodes, as they are called, eventually took on independent life as homeopathic medicines and are still used today in their own right, even for patients who have never suffered from the relevant diseases. The origin of some of the nosodes is obscure. (Strictly speaking, when actual tissues are used in this way they should be called sarcodes rather than nosodes but the difference is not important.)

Many of the older nosodes developed by Hering and his contemporaries have been forgotten, which is hardly surprising in view of the somewhat implausible claims that Hering made for them. In 1830, for example, we find him recommending that farmers eradicate weeds by means of their potentised seeds and that lice be removed by means of a 30th potency of their own relatives. (Dudgeon, who reports this, mischievously implies that this is to be administered individually to the lice.) And some American homeopaths went even further. We read of one who suffered an upset stomach after eating a particular pudding and accordingly solemnly potentised the pudding.

The last and in many ways the most influential American homeopath in the nineteenth century was James Tyler Kent, who took the ideas of the high-potency school to their limit.

Chapter 8

Kentian Homeopathy

In the opinion of his pupils and followers, James Tyler Kent (1849–1916) is second in importance only to Hahnemann himself, and perhaps not even second.

> His intense desire to alleviate suffering, to eradicate disease, led him to concentrate, by the power of his indomitable will, the forces of his vast intellect. He gave himself unstintingly to the arduous task of acquiring that deep knowledge by which he scaled the heights of the Homeopathic Law of Cure. Here his unclouded vision beheld the genius of Samuel Hahnemann. He grasped the Master's thought, he wielded the healing power, he reached greater [sic] heights.

This hyperbolic passage is from an obituary published in America in 1917, and its tone of near-adulation is by no means exceptional. Another writer, for example, describes Kent as "one of the greatest masters in medicine the world has ever known", while yet another says that "since Hahnemann only in this one man have been so brilliantly combined the three attributes that enable Homeopathy to stand so firmly in these times of medical Nihilism".

Who, then, was this remarkable medical paragon?

Kent originally qualified as what was known as an Eclectic physician, but his outlook was changed when his wife became ill and begged Kent to place her under the care of a homeopath. Although he had no faith in this system of medicine he acceded to his wife's request and sent for one Dr Phelan, who effected a dramatic cure. This happy event led to Kent's conversion to homeopathy. He made rapid progress in his study of the subject and in 1882 was appointed to the Missouri Homeopathic College as (rather oddly) Professor of Surgery; in 1889 he joined the staff of the Philadelphia Postgraduate School of Homeopathy. After his first wife's death he married a homeopathic physician, who cooperated with him in the writing of his three major works: the *Lectures on Homeopathic Philosophy*, the *Lectures on Homeopathic Materia Medica*, and the *Repertory to the Homeopathic Materia Medica*, the book on which his reputation principally rests today. So pre-eminent has Kent's *Repertory* become that, although numerous other repertories exist, Kent's is usually referred to as *The* Repertory, as if there were no other.

Photographs of Kent show him wearing a disapproving expression and an unkempt moustache. Both these features are probably significant, for an obituary by a Dr Minerva Green remarks on his "ill-fitting and ill-assorted clothes", and says that he was "a sensitive, embittered, retiring man in later years as he thought one after another did him wrong". As Dr Green remarks, this last trait reminds one of Hahnemann, who also suffered from a feeling of persecution in his later years.

Figure 8.1: J.T.Kent

In 1900 Kent became Dean of Dunham Homeopathic College in Chicago. In 1908 all the homeopathic colleges in Chicago were merged into the Hering Homeopathic Medical College, of which Kent was president until 1911. In that year the Government closed the Hering College, together with many other homeopathic colleges throughout the USA, on the grounds that they were not up to medical standard. This was the end of Kent's medical career.

The Eclectic school of medicine in which Kent began his career needs a few words of explanation. The founder of this school was Dr Wooster Beach, although he eventually broke away from his own followers. He preached moderation in the use of conventional therapy and introduced several native American remedies, as did a somewhat later "medical reformer", C.S. Rafinesque, who probably inspired the use of the term "eclectic" to describe the new movement. Both Beach and Rafinesque were influenced by contact with so-called "root and Indian" doctors, as indeed was the homeopath Edwin Hale, whose "New Remedies" were later incorporated into the homeopathic materia medica, often without the formality of a standard homeopathic proving.

Kent's apprenticeship in Eclecticism was due to a later and very influential teacher, John M. Scudder. In his recent book *The Magical Staff: The Vitalist Tradition In Western Medicine*, Matthew Wood makes a convincing case for his view that many of Kent's ideas, even after his conversion to homeopathy, derived from Scudder, who was Kent's professor of pathology and philosophy of medicine in 1870. Although Wood fully recognises the central importance that Swedenborgianism came to have for Kent, he believes that many of Kent's most characteristic teachings come from Eclecticism. Hahnemann, Wood says, had a basically scientific, objective view of the role of the physician, whereas Scudder taught that the life force of the physician was an important part of the healing process. "Unfortunately, [Kent] was unaware of the fact that he was borrowing the central tenet of Eclecticism, presenting it as a Hahnemannian approach, when it was really a homeopathic heresy."

The picture of Kent that emerges from Wood's study differs in several ways from that we have been accustomed to hitherto. It seems that Kent was regarded as something of an interloper by the homeopaths of his day, and his views were by no means universally welcomed. Wood describes him as appearing suddenly on the homeopathic scene in about 1885, and he quotes Julian Winston, editor of *Homeopathy Today*, as saying that Kent "rode out of the West like the man in the black hat". This swashbuckling version of Kent is perhaps a little difficult to take in, but it is probably broadly correct.

In his published writings Kent made few direct references to Swedenborg's influence on him but he did freely acknowledge it to his pupils, claiming that the teachings of Swedenborg and Hahnemann corresponded perfectly. Although this is an overstatement it is true that there is much common ground between Swedenborg and the later Hahnemann, which is what interested Kent.

The best place to gain an insight into Kent's thought is his very influential *Lectures on Homeopathic Philosophy*. These are cast in the form of extended commentaries on Hahnemann's *Organon* (the fifth edition; fortunately for Kent's peace of mind he died before the discovery of the sixth edition, in which Hahnemann contradicts some of the main ideas in the fifth). Kent's method is to take a passage from *The Organon* and dilate upon it, much in the manner of a preacher making use of a text from Scripture.

Kent is a curious writer. His manner is frequently hectoring and sometimes downright abusive — an unattractive trait, perhaps imitated from Hahnemann. Thus, of the wretched pseudo-homeopath who stoops so low as merely to remove symptoms instead of eradicating their cause (for example, by giving morphine to a patient with a kidney stone instead of keeping him waiting for an hour or two while looking for a suitable homeopathic remedy) he writes: "What a simple-minded creature he must be! What a groveller in muck and mire he must be, when he can meditate upon such things, even a moment." Here speaks the true fanatic.

Kent's own moral standards are made universally applicable. If patients use contraception, we learn, it will not be possible to cure their chronic diseases. "The meddling with these vices and the advocating of them will prevent the father and mother from being cured of their chronic diseases. Unless people lead an orderly life they will not be cured of their chronic diseases. It is your duty as physicians to inculcate such principles among them that they may lead an orderly life. The physician who does not know what order is ought not to be trusted."

Kent obviously felt strongly on this subject, as is evident from the frequency with which he repeats himself, driving the point home in case we have missed it. The effrontery of the passage is breathtaking, but unfortunately quite typical of the man. Its uncompromising assertiveness is also typical. No evidence is offered for the remarkable pronouncement that chronic disease is incurable in those who practise contraception; we must simply accept it on Kent's authority.

"Authority" is in fact a key word with Kent. Notice the emphasis on order, always beloved of authoritarians. Elsewhere he writes: "It is law that governs the world and not matters of opinion or hypothesis. We must begin by having a respect for law, for we have no starting point unless we base our propositions on law. So long as we recognise men's statements we are in a state of change, for men and hypotheses change. Let us acknowledge the authority."

But whose authority are we to acknowledge? Presumably Hahnemann's; but surely Hahnemann was a man, and therefore no more exempt from error than other men? Not so, Kent implies, for Hahnemann had discovered a divinely ordained law. Homeopathy is an inspired science, which is the only true kind of science; all the rest is mere opinion. It is therefore not merely foolish but actually impious to question Hahnemann. By implication it is also impious to question Kent.

This invincible belief in his own rightness pervades everything Kent wrote. All his statements are made *ex cathedra*; nowhere does he express the faintest doubt about anything, nowhere does he offer any evidence in support of what he says; everything has to be taken on trust. It is, as he accurately remarks, a matter of acknowledging the authority.

Now, whatever one's assessment of Kent's status as an "authority", there is no denying that his procedure is the very reverse of scientific. For the scientific method consists essentially in a willingness to question authority and not take things for granted. The development of science in Europe from the seventeenth century onwards depended largely on the fact that people were beginning to question traditional ideas, especially the authority of Aristotle, whose writings had been regarded as the ultimate court of appeal for over a thousand years. Reverence for authority is incompatible with science. Kent is therefore deeply anti-scientific, and his version of homeopathy is founded on metaphysics.

He himself is quite frank about this. "In all your experience, even if you live to be very old," he writes, "you will find a very poor lot of homeopaths among those who do not recognise Divine Order. You will find among them false science and experimentation, but never any government of principle, no thought of purpose, order or use."

Kent's belief that homeopathy is founded on divine order and that disease results from transgression of this order pervades his writings but nowhere does it emerge more clearly than in his discussion of psora, which he regards as a moral as well as a physical contagion affecting all mankind.

> The human race walking the face of the earth is little better than a moral leper. Such is the state of the human mind at the present day. To put it another way, everyone is psoric ... A new contagion comes with every child.

Psora is the root of all evil and the other chronic miasms, sycosis and syphilis, are secondary to it.

> The human race becomes increasingly sensitive generation after generation to this internal state [psora], and this internal state is the underlying cause which predisposes man to syphilis. If he had not psora he could not take syphilis; there would be no ground in his economy upon which it would thrive and develop.

Kent's interpretation of the psora doctrine is uncompromisingly metaphysical or spiritual. Psora results, he says, from a disorder at the inmost level of thinking, willing, and acting — the three functions of mind in the Swedenborgian scheme. As a consequence, Kent places the main emphasis in medicine selection on the patient's mental symptoms. Hahnemann, it is true, regarded the psychological aspects of disease as very important, but Kent took this trend much further. He devoted nearly 100 pages of his *Repertory* to Mind, compared with a mere nine in von Boenninghausen's.

Kent and potencies

In view of Kent's deep belief in the more extreme aspects of Hahnemann's thought it is no surprise to find him an enthusiastic advocate of ultra-high potencies. He would have no truck with anything lower than a 30th centesimal but this was for him merely the beginning of the scale, and his practice soared into the dizziest heights

— the 1000th centesimal (usually written M), the 10,000th (10M), 100,000th (CM), and even the millionth (MM) being commonly used by Kentians.

Kent is forced to acknowledge that in this respect he has gone beyond the Master. Hahnemann used and advocated the 30th and occasionally toyed with the 300th, but he went no higher. He also maintained that these high potencies caused more transient aggravations. Kent, however, claimed that the very highest potencies (CM and MM) were extremely powerful and if given incautiously could cause very serious aggravations or even kill the patient. When in doubt, therefore, the Kentian prescriber should give a "low or moderately low" potency (30c or 200c).

All this talk of high potencies begs an important practical question: namely, are Kent's potencies really what they claim to be? Even in Hahnemann's lifetime machines were invented to make potencies, but Hahnemann did not take them very seriously. When the Americans began to think in terms of CM and MM potencies, however, it was obviously impossible for them to make them by hand in the Hahnemannian manner.

A quick calculation will show why. To make a single centesimal dilution by Hahnemann's technique requires, say, 100 ml of water and takes 3 minutes. To make a 30th centesimal dilution therefore requires 3 litres of water, 30 sterile bottles, and takes one and a half hours, which is acceptable. To make a 1000th centesimal (1M) dilution would require 100 litres and 1000 sterile bottles and would take 50 hours' work. A CM dilution would require 10,000 litres of water, 100,000 sterile bottles, and would take over 200 days with relays of pharmacists working round the clock. Clearly we are here in the realm of fantasy.

An edition of Kent's *Lectures*, published in 1919, contains an advertisement by a firm of manufacturing homeopathic pharmacists, Erhart and Karl of Chicago. This firm claims to have 900 remedies made by hand to the 1000th potency. From this point on, "Kent Potencies" are supplied. These take the hand-made 1000th potencies as starting point and allegedly raise them further by mechanical means, using a machine invented by Kent. Even higher potencies were made by another machine invented by Dr H.C. Allen. This begins where the Kent machine leaves off, using the Kentian CM potency as a starting point.

We need to be clear about what is being claimed here: that the effects of hand potentisation can be imitated by machines of various kinds that work on a quite different principle. Potentisation is allegedly achieved by allowing a continuous stream of water to pass through a tube (a circular one in the case of the Allen machine); the swirling motion, produced by a propeller, is supposed to reproduce the effect of Hahnemannian succussion. (For a good review of the history of these machines, see Julian Winston's article — Appendix E.)

Now, even if we grant for the sake of argument that Hahnemann's dynamisation is a real phenomenon, what guarantee or even likelihood is there that the Kent and Allen machines lead to the same result? It is quite characteristic of Kent that he is totally unconcerned with questions such as these — in fact he never considers them. He takes his stand on a principle and that is enough for him — and, he implies, so it should be for us.

Kent's materia medica

Important and influential though Kent's philosophical ideas became for homeopathy, it was probably his treatment of the materia medica that did most to attract students to sit at his feet. The principal difficulty faced by newcomers to homeopathy was the shapelessness of the material they had to master. Hering had introduced the idea of giving the medicines a personality, as it were — to dramatise them. Instead of presenting students with long lists of unconnected facts he painted word-pictures of the kind of patient who was supposed to need the medicine in question. Hence we have sulphur as the "ragged philosopher" (untidy, absent-minded), while arsenicum is the opposite (fussy, tidy — the "gold-headed cane" patient). This way of describing the medicines was adopted by Kent. In his book we read that sepia, for example, "is suited to tall slim women with narrow pelvis and lax fibres and muscles; such a woman is not well built as a woman". (This probably tells us as much about Kent as it does about sepia; one pictures both Mrs Kents as stout buxom blondes.)

This method of presenting the medicines was undoubtedly much easier for students to assimilate, and Kent's lectures, if verbose, were certainly more readable than the standard reference works. However his approach involved a considerable dilution of the original similia idea.

For one thing, a lot of the material in Kent's descriptions was "clinical", being derived (presumably) from Kent's own observations in patients. Certainly much of it could not have come from provings; it could hardly be claimed, for example, that sulphur can make someone untidy who is not so already. For another, Kent, like the sorcerer's apprentice, had started a trend he probably did not intend and could not control. Although he advised his students to read the original provings it is difficult to avoid the suspicion that few of them did so; later generations of Kentian homeopaths, at any rate, came more and more to rely on the writings of Kent himself, and this brought about a new attitude to the materia medica.

Thanks to Kent the trend towards thinking about the remedies almost as if they were personalities in themselves became deeply embedded in homeopathy and it led, as we see, to a trend towards "constitutional prescribing".

The significance of Kentianism

In Kent's own day his views were approved by only a minority of American homeopaths and it may seem surprising that I have given them so much space. In later years, however, they were to become remarkably influential among homeopaths outside America, as I shall explain in the next chapter. Kentian homeopathy represents Hahnemann's later, more extreme, ideas taken to their logical limit and furnished with a Swedenborgian underpinning. Its principal features could be summarised as follows:

- Insistence on the theoretical aspects of Hahnemann's thought, especially the miasm doctrine and vitalism.

- A corresponding rejection of modern scientific and pathological knowledge as a guide to prescribing.

- Great emphasis on the importance of psychological symptoms in prescribing.

- Insistence on the use of very high potencies.

All these features naturally widened the gap separating homeopathy from orthodox medicine. This didn't worry Kent or his disciples — indeed they rejoiced in it — but it was to have a profound effect on the character of later homeopathy.

Chapter 9

The Twentieth Century

In the early years of the twentieth century an English homeopathic doctor, Margaret Tyler, went to America to study under Kent. On her return full of enthusiasm for the new teaching she published a pamphlet in which she criticised the prevailing orthodoxy. I have not managed to find a copy of this pamphlet but it is referred to by other homeopaths of the time and it apparently led to Tyler's ceasing to attend meetings of the British Homeopathic Society for a couple of years, so tempers were evidently frayed.

In spite of this friction, Tyler was unrepentant and in 1907, in conjunction with her mother, Lady Tyler, she instituted a scholarship to send doctors to the USA. An early beneficiary was Dr (later Sir) John Weir, who soon after his return to England in 1909 was appointed Compton Burnett Professor and Honorary Secretary of the British Homeopathic Society. Under the influence of Drs Tyler and Weir in London and Dr Gibson Miller in Glasgow, Kentianism made rapid progress towards becoming the prevailing homeopathic orthodoxy in Britain. The Hughesian Old Guard naturally resisted the new trend, but though they remained unconverted they were ageing, and by the end of the First World War opposition to Kentianism had all but ceased.

The result was a decisive shift in Britain away from scientific homeopathy towards the more extreme form taught by Kent. So the gap separating homeopaths from orthodox doctors grew wider, in spite of some attempts to bridge it. But the tradition of royal patronage of homeopathy continued, and when the National Health

Service was set up in 1948 the homeopathic hospitals were included. This ensured the survival of homeopathy within the British medical scene. In 1950 the British Homeopathic Society became the Faculty of Homeopathy, established by Act of Parliament. It is now linked with the British Homeopathic Association, a charity founded in 1902 to promote education and research in homeopathy.

This remarkable degree of official recognition makes Britain unique in the world; only India comes anywhere near it in this respect. As a result, doctors came to study in Britain from all over the world. Though not all the leading homeopaths of the twentieth century have been Kentians — one of the best known, Dr Charles Wheeler, had reservations about Kent and maintained a good deal of respect for Hughes — the prevailing orthodoxy was emphatically Kentian and newcomers were given to understand that Kent's version of homeopathy was the purest and most authoritative. This was how I encountered it myself in 1974.

Even so, British homeopaths tended to be less extreme than Kent had been. They preferred high potencies (the highest being obtained from America, where they continued to be made on machines) but many used low potencies as well on occasion. It was quite common to give a single dose of a high potency on "constitutional" lines and to supplement this with a daily dose of a low potency to take account of particular symptoms. Vitalism was not a central question in Britain and there was little discussion of metaphysical issues in British homeopathic circles; indeed the influence of Swedenborg on American homeopathy was probably unknown to many British homeopaths. As for relations between homeopathy and orthodox medicine, many British homeopaths would have liked to heal the breach but their attempts to do so were unsuccessful, largely owing to the hostility of orthodox doctors.

Probably the most important effect of the Kentian influence was the way in which the materia medica was taught. No longer were students expected to read the original provings as they had been in Hughes's day, and it seems unlikely that more than a tiny minority did so. Kent's writings were now the authoritative source, but not the only source. Margaret Tyler also tried her hand at the art of painting word-pictures of medicines and soon outdid her mentor in readability and verve. Before long these "remedy pictures" became with her almost Dickensian. Here, for example, is her description of sepia.

> Sepia has been called the washerwoman's remedy and not without cause. Picture her — the sallow tired mother of a large family, on "washing day". She is perspiring profusely: pouring under the arms. She cannot be shut in, because of the heat and the stuffiness which make her feel faint — yet the cold wind that rushes in at the open window is almost unbearable. Her back aches fearfully
> . . .
>
> The worry of the children is more than she can bear. Her baby wants to be picked up and carried, and wails and screams. The quarrels of the penultimate babies, engaged in scratching each other's eyes out, are more than she can bear. And when her 6-year-old starts drumming with a spoon on a pot, she can stand it no more. She snatches the tin pot and hurls it away, and smacks her small son; which does not improve matters. He howls dismally *and she does not care . . .*

And so on. We are here a world away from the austerity of densely packed narratives of provings in Hughes's *Cyclopaedia*, let alone from the bare symptom lists in Hahnemann's *Materia Medica Pura*. Even Kent seems dry and restrained in comparison. In the hands of Margaret Tyler and her colleagues homeopathy gave up any pretence of being scientific and became, for better or worse, more like an art form.

It was in this guise that it was taken up by a growing body of non-medically-qualified practitioners. There had always been a tradition of lay practice in homeopathy (rather as happened in psychoanalysis); Melanie, Hahnemann's second wife, practised as a homeopath and von Boenninghausen, one of the most influential of Hahnemann's early disciples, had been a lawyer. If homeopathy had developed on the lines advocated by Hughes the lay practitioners would have been squeezed out, since Hughes's approach depended on a knowledge of physiology and pathology. But the writings of Margaret Tyler and her colleagues made homeopathy accessible to people who lacked a medical background — hence their continuing popularity today.

Constitutional prescribing

Although it is difficult to be certain, it seems likely that the concept of "constitutional prescribing" in the modern sense is largely due to Margaret Tyler and her associates. Many people today think that this is completely central to homeopathy but it is in fact a relatively recent development.

The idea is that, at least in treating chronic disease, the homeopath's aim should be to "find the patient's constitution". If the right constitution is identified, so the theory goes, giving the relevant remedy will bring about a cure more or less regardless of the actual symptoms the patient is complaining of. For example, suppose a patient comes to the homeopath with asthma. The homeopath may spend time asking about the things that make the asthma worse or better (changes in weather, time of day, different foods and so on), but also, if she is a constitutional prescriber, she will ask a lot of questions that have nothing ostensibly to do with the asthma. They concern such things as fears, moods, food likes and dislikes, and weather preferences in general. These are the so-called "Mentals" and "Generals", discovering which forms the most important part of "taking the case" in the Kentian system.

The same notion can be applied to animals. According to Francis Hunter, writing in The Faculty of Homeopathy newsletter *Simile* (January 2008), the arsenicum cat, for example, loves warmth and is neat and tidy, spending plenty of time grooming. Sulphur cats, on the other hand, "tend to be indolent, scruffy moggies who prefer to be outside, do not think too much about grooming, are inclined to be loners and are not averse to the occasional scrap". But most cats are of natrum muriaticum constitution; they expect their owners to attend to their every whim but don't like too much fussing or stroking. They can be irritable and even bad-tempered at times. They keep away from strangers and when stroked may well turn round and nip after a short time.

The constitution idea has a certain charm, but contrary to what quite a few homeopaths believe, it does not originate with Hahnemann. The closest he comes to it is in a couple of glancing references. He suggests that pulsatilla is best suited to gentle weepy girls, and that nux vomica is often needed by over-indulged business men. That is all he says on the matter. It was Hering and later Kent in the USA who produced the first portraits of arsenicum, sulphur, and other remedies in terms of characters (or caricatures); without these remedy pictures constitutional prescribing could hardly have developed. But it was not the intended consequence.

Kent used remedy pictures as a teaching aid but he explicitly condemned the practice of constitutional prescribing, saying (correctly) that it was unhomeopathic, because homeopathy is concerned with changes from the normal whereas constitution is normal. But Margaret Tyler, John Weir, Gibson Miller, and other British homeopaths in the first half of the twentieth century ignored these warnings and made the quest for constitution the prime task of the homeopath, though admittedly an exact constitutional match cannot be found in every case. Still, it remains the ideal.

We often hear claims that homeopathy treats the patient as an individual. However, this is true only within rather narrow limits. The constitutions are closely related to certain remedies called "polychrests" — those that are associated with a large number of symptoms (there is no precise definition of a polychrest). They include such well-known remedies as sulphur, silicea, phosphorus, lycopodium, sepia and pulsatilla. In theory there are many hundreds of homeopathic remedies that might be chosen, but in practice far fewer are used to any great extent (though some homeopaths take a pride in seeking out and using unusual remedies).

It is fairly obvious that if a medicine is associated with many symptoms this medicine will turn up frequently when a patient's symptoms are looked up in a repertory, and this is what happens. A homeopathic consultation usually yields one of the polychrests, and probably most homeopaths use only about 20 or 30 remedies with any frequency, Hence it could be argued that homeopathy is really no less stereotyped than conventional medicine, but it is stereotyped in a different way.

Clinical studies

The great majority of homeopathic writing in the first half of the twentieth century was "anecdotal", consisting largely of case studies. It would be unfair to criticise homeopaths for this, since the same was true at this time for orthodox medicine as well. But after the Second World War British doctors began what was in effect a medical revolution, by introducing the concept of clinical trials. That is, they began to study the effectiveness of treatment by applying statistical methods. Today the standard way of doing this is by a randomised controlled trial (RCT). The vogue for RCTs was slow to catch on within homeopathy, which remained a medical backwater until quite recently. Many homeopaths raised objections in principle to applying statistical methods in homeopathic research, claiming that the need to individualise prescriptions made it impossible to allocate patients to treatment groups. Nevertheless, ways of doing this were found, and although it has not pleased everyone the trend towards "evidence-based medicine" means that the process is irreversible. I return to this question in Chapter 11.

Scientific studies

Although the prevailing homeopathic climate in the early twentieth century was Kentian it would be wrong to suggest that British homeopaths had cut themselves off from science entirely. A minority did carry out research, some of which resulted in new kinds of homeopathic medicines. The best known of these are the "bowel nosodes" derived from bowel bacteria and developed by Paterson and Bach. (Bach was a pathologist at the London Homeopathic Hospital, who later abandoned science when he went on to discover his "flower remedies".)

The most important scientific homeopath of the first half of the century was Dr William Boyd of Glasgow. He had considerable technical and engineering expertise as well as a scientific cast of mind and did some experimental work on the potency question which seemed to give positive results. His most interesting project, however, was the Emanometer.

The inspiration for this research was the work of an American doctor (not a homeopath) called Abrams, who in turn had derived the idea from Dr Stern White of Los Angeles in 1914. Abrams invented a machine called a Reflexophone, which he claimed detected "energy fields" affecting patients. Purchasers of the Reflexophone were not supposed to open it, but Boyd, obeying the letter but not the spirit of this injunction, x-rayed it and found that it could not possibly do what Abrams claimed. Although he was sceptical of Abram's methods Boyd felt that there was something genuine at the bottom of it all and he therefore set to work to design his own machine, the Emanometer, which was quite different.

Boyd started this work in the early days of wireless, and probably for this reason the Emanometer has a certain resemblance to a crystal set, the forerunner of the modern radio. It was more complicated, however, and Boyd was careful to insist that the "energy" it detected was not necessarily identical with radio waves.

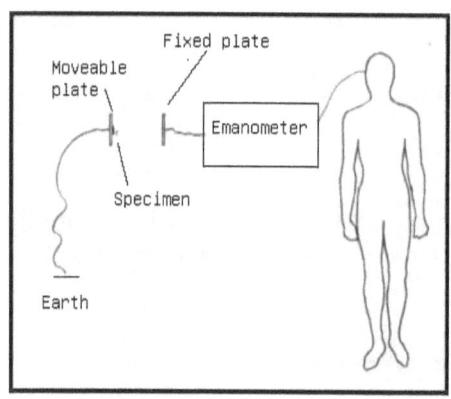

Figure 9.1: The Emanometer

Although the details of the Emanometer design were rather complicated the basic set-up was quite simple. The specimen to be tested — blood, tissue, or a homeopathic medicine — was attached to an earthed plate, which was set at a variable distance from a second (fixed) plate. The fixed plate was connected to the circuitry of the apparatus, which was in turn connected to the forehead of a person (usually a boy) who acted as a detector (Fig. 9.1).

To carry out the test Boyd would percuss (tap) the detector's abdomen in the way that a doctor percusses a chest. By so doing he would map out areas of relative dullness, which he recorded together with the settings he was using on his machine. He would then insert a specimen (say, blood from a patient) and see what effect this had on the areas of dullness and on the machine settings. He also tested homeopathic medicines to see how they changed the reading.

This is the merest outline of Boyd's very painstaking method. He spent many years trying to perfect his technique and designing improved versions of the Emanometer. Though always commendably cautious about his results he became convinced that he was discovering something. He was apparently able to detect abnormalities in patients with fair accuracy, sometimes before the patients themselves were aware that anything was wrong, and he could also distinguish different medicines and potencies. On the basis of this research he built up an Emanometer classification of homeopathic medicines that was used by some homeopaths of the time.

In 1924 a committee under an eminent physicist, Lord Horder, investigated the Emanometer. Later the committee was joined by E. J. Dingwall, research officer of the Society for Psychical Research, who was an authority on fraud. After exhaustive tests the committee concluded that

> certain substances, when placed in proper relationship to the Emanometer of Boyd, produce beyond any reasonable doubt changes in the abdominal wall of the subject of a kind which may be detected by percussion ... The phenomena appear to be extremely elusive and highly susceptible to interference [and] it would be premature at the present time even to hazard in the most tentative manner any hypothesis as to the physical basis of the phenomena here described.

In other words, Lord Horder and his committee were sure that Boyd could detect *something* with his apparatus but they had no idea what it was or what it meant. They were also careful to say that there was as yet no good evidence that the Emanometer could be used in diagnosis or treatment — a cautious attitude that Boyd fully shared.

The main weakness of the Emanometer was the need to use a human subject as a detector of the mysterious energy. In spite of many years' hard work Boyd never succeeded in eliminating the need for this detector and he died with most of the secrets of the Emanometer undiscovered. After his death his sons (one a physiologist, the other a homeopathic physician) tried to continue his work but without success, and no one else has taken it up.

Today other machines purporting to allow the selection of homeopathic medicines by "energy detection" are marketed but they are unsupported by research approaching anywhere near to Boyd's in quality. Some homeopaths use pendulums and allied "radionic" techniques to help them choose medicines, but again serious scientific evidence for their claims is almost wholly lacking.

Boyd's Emanometer research is tantalising but ultimately baffling. It is in many ways reminiscent of much research in parapsychology, which likewise seems constantly to promise to yield firm evidence and then fails to do so. This is probably not accidental. There are numerous parallels between alternative medicine, especially homeopathy, and the paranormal in respect of how attempts to verify them scientifically have developed. Research in the paranormal has gone on for over a hundred years but there is still no agreement about what, if anything, has been achieved. A major difficulty faced by such researchers is the lack of any adequate theory to explain how the results they claim to have found could conceivably be produced. This, probably more than anything else, is the reason that their work has not been more widely accepted as valid. The alleged results are so difficult to accommodate within the prevailing scientific world view that most critics feel instinctively that there simply must be something wrong with the research, no matter how carefully it seems to have been done.

Critics of homeopathy voice very similar objections, and it seems unlikely that homeopathy will be taken seriously by many scientists unless and until a plausible mechanism for its effects can be suggested. If you find the evidence for parapsychology and psychic phenomena to be convincing you will have little difficulty in accepting Boyd's results, since they belong to the same world of mysterious and elusive phenomena. If, on the other hand, you are unconvinced by this research, you will probably reject Boyd's findings as well.

These rather dubious considerations lead us naturally into the subject of the next chapter, where I look at some other interactions between homeopathy, the occult, and Mesmerism.

Chapter 10

The Occult and Mesmer

By linking homeopathy with Swedenborgianism the American high-potency school established a connection with occultism, but this is not the only one of its kind that exists. There was a counterpoint of occultism running through homeopathy right from the beginning.

We may conveniently begin this rather obscure story by looking at some of the resemblances that exist between Hahnemann's ideas and those of the sixteenth-century physician Theophrastus von Hohenheim, commonly known as Paracelsus, who came from the alchemical tradition. He was an extreme individualist, who made himself unpopular by attacking the orthodox physicians of his day as incompetent bunglers. This of course at once puts one in mind of Hahnemann, and so does the fact that Paracelsus was on bad terms with the apothecaries because he prescribed single medicines instead of the customary complex mixtures.

These resemblances are comparatively superficial. More significant are Paracelsus's ideas about disease and its treatment. Like Hahnemann, he rejected the idea that disease could be reduced to a certain number of diagnostic categories, insisting that each patient must be treated as an individual. And he recognised a version of the similia principle, saying that "likes must be driven out by likes", though his meaning is not exactly the same as Hahnemann's.

What Paracelsus taught was a version of the ancient "doctrine of signatures". It seems to be a world-wide belief, based on magic, that natural substances, especially herbs, reveal by their appearance or other characteristics the medicinal purposes for which they are suited. Thus yellow plants are good for treating jaundice, for example. Paracelsus adopted this idea and gave it an anatomical twist, claiming, for instance, that plants with heart-shaped leaves could be used to treat heart disease. Although this is not homeopathy it certainly reminds one of it.

Another way in which Paracelsus anticipated Hahnemann was in his habit of giving drugs in minute doses. A very small dose of medicine, he said, could overcome a great disease, just as a small spark can ignite a great heap of wood. Moreover, the medicinal power of drugs was for Paracelsus, as for Hahnemann, a spiritual thing, which could in principle be separated from the crude substance. In this there was a clear echo of alchemy, for a tiny fragment of the philosophers' stone was held to be enough to transmute a large amount of base metal — or to cure any disease.

The numerous parallels between Hahnemann and Paracelsus present us with a puzzle. It is difficult to think that they are due to chance, especially in view of the fact that Hahnemann read so widely. It seems unlikely that he would not have come across Paracelsus's ideas in books or through his Masonic contacts, for early nineteenth-century German Masonry was influenced by ideas of this kind via its connections with Rosicrucianism.

Yet Hahnemann nowhere refers to Paracelsus by name and he has merely one disparaging reference, in a footnote, to the "childish" doctrine of signatures. It seems that late in his life one of his followers did draw his attention to the similarities between his ideas and those of Paracelsus, but Hahnemann replied that he had never heard of him.

This may of course be an example of Freudian "forgetting". In any case, among post-Hahnemannian homeopaths some were deeply influenced by the occult alchemical tradition to which Paracelsus belonged, and these homeopaths did not hesitate to make the connection explicit.

The Golden Dawn

Probably the earliest manifestation of this is provided by the Hermetic Order of the Golden Dawn, the magical society which included among its members not only the poet W.B. Yeats but also a number of homeopathic doctors. The Golden Dawn had a medical flavour from its inception, for it was founded in 1888 by Dr Wynn Westcott, a physician turned coroner. For this purpose Westcott forged letters of authorisation from a certain "Fraulein Sprengel", an eminent Rosicrucian adept who he said lived in Germany. Westcott invited another doctor, W.R. Woodman, and a strange occultist called S.L. MacGregor Mathers to join him as Chiefs of the Order.

The Rosicrucian tradition on which the Golden Dawn was allegedly based had itself strong links with medicine as well as with alchemy and also with Paracelsus. It derived from the publication in Germany, in the early seventeenth century, of the "Rosicrucian Manifestos". These mysterious texts, supposedly written by a secret Brotherhood of initiates, caused a tremendous furore in Europe when they first appeared and their effects were felt in all kinds of unlikely places. Francis Bacon, for example, appears to have known about them, and Isaac Newton likewise; while the idea of a secret brotherhood of savants probably inspired Robert Boyle and other founders of the Royal Society.

The Manifestos described the life and career of the supposed founder of the Order, Christian Rosenkreutz. He was said to have been a German monk who travelled to the East and there acquired much esoteric alchemical and medical knowledge. On his return he instituted the Brotherhood to preserve this knowledge. He was buried in a secret vault, which contained all the books written by him and his colleagues and — a significant inclusion — one by Paracelsus, who though not a member of the Order was claimed as a kind of fellow-traveller. The vault was intended to be a time-capsule to preserve all this knowledge, and it was the accidental rediscovery of the vault, whose location had been forgotten, that was said to have prompted the publication of the Manifestos.

The members of the Golden Dawn believed in the literal truth of the Rosenkreutz legend and went so far as to reconstruct a replica of the vault in which to perform their magical rites. Christian Rosenkreutz himself was supposed to be a physician and his followers were expected to support themselves by practising medicine. In view of this, and the association with Paracelsus, it is easy to understand why Rosicrucianism should have attracted doctors who were drawn by their temperament towards the occult. Fourteen medical men, in addition to Westcott and Woodman, were members of the Golden Dawn before 1900, and many of these were interested in homeopathy. One of the most prominent members, Dr Edward Berridge, was a well-known homeopathic doctor who wrote a book on homeopathy and whose name appears as a prover in the American homeopathic literature at this time.

When it became clear that the authorisation for setting up the Golden Dawn that Westcott had obtained from "Fraulein Sprengel" was bogus the Order broke up in confusion. But one medical member, Dr R.W. Felkin, refused to be discouraged. There must exist somewhere, he supposed, Secret Chiefs, guardians of esoteric knowledge, if only they could be found, and shortly before the First World War he set off on a series of travels in Germany to look for them. This quest led him to Rudolf Steiner, the mystical philosopher who founded Anthroposophy. Felkin apparently hoped that Steiner would appoint him as his representative in England, but in this he was disappointed, and Steiner does not seem to have taken him very seriously.

Steiner himself, however, took a great deal of interest in medicine, and later developed a therapeutic system that is in many ways a refinement of Paracelsus's ideas. It also has a good deal in common with homeopathy and continues to attract some homeopathic doctors.

Anthroposophical medicine

Though not himself qualified in medicine, Steiner attracted a number of physicians to him and towards the end of his life he lectured extensively on medicine. In 1921 Ita Wegman came into contact with Steiner, and with his encouragement began her medical training in Switzerland. After qualifying she founded the Clinical–Therapeutic Institute at Arlesheim in Switzerland, where Anthroposophical methods of treatment are still in use today. In addition a laboratory was set up at Dornach for the investigation and production of Steiner's remedies, and this work later gave rise to a number of commercial manufacturing companies in different countries.

Steiner's medical ideas are rather similar to those of Hahnemann though they also derive from earlier sources, especially Paracelsus and the alchemists, and Steiner emphasised symbolism and occultism, including the doctrine of signatures. But many Anthroposophical medicines are the same as those used in homeopathy though they are often given as mixtures instead of singly. The Hahnemannian method of potentisation is sometimes used but Steiner also invented some more complicated procedures.

For example, metals are often "vegetabilised" by passage through a plant. A metal is added to the soil in which the plant is growing; next year the plant is composted and used to fertilise a second generation of plants, and the process is repeated for a third year. This is said to dynamise the metal very effectively, while the influence of the metal causes the plants to direct their action to a particular organ or system.

There has long been an uneasy tension between homeopaths who wish to make their subject wholly scientific and respectable and those who have leanings towards the mystical or the occult. (Homeopaths with strong Christian beliefs have also objected to the occult element, though for different reasons.) Today, naturally, the scientifically minded are in the ascendant; the talk is all of evidence-based medicine, double-blind trials, and the physics of water molecules. Yet there has always been a movement within homeopathy in the opposite direction, and some still use Anthroposophical medicines alongside homeopathy to a greater or lesser extent.

There is a blurring here of the boundary between science and magic, something that has always characterised homeopathy since its inception. Even today, some homeopaths are drawn towards unconventional and unscientific means of selecting remedies, such as dowsing. In this, as in other respects, homeopathy harks back to its origins. We tend to think of Hahnemann as a nineteenth-century figure, but we forget that his formative years were spent in the eighteenth century. We don't need to go much further back than that to reach a time when doctors routinely used astrology to help them make their diagnoses.

Our modern sciences had their origin in less reputable activities: astrology fathered astronomy, alchemy chemistry. Isaac Newton spent many years in the practical pursuit of alchemy; Kepler, who formulated the idea that the planets move in ellipses rather than circles, was motivated by the desire to prove that the orbits of the planets correspond to the Platonic regular solids. In the seventeenth century mathematics was only just ceasing to be thought of as a form of magic. Modern medicine, too, developed painfully and slowly from less "rational" sources. For at least some homeopaths, an important part of the appeal of their system of treatment is that it is closer to the realm of magic. (See Chapter 12.)

Mesmer and Hahnemann

No discussion of the occult element in homeopathy would be complete without a reference to Anton Mesmer and his "animal magnetism". There are considerable similarities between the ideas and even the careers of Hahnemann and Mesmer, and Hahnemann made complimentary references to Mesmer in his later writings, as did a number of the early homeopaths. It is therefore worth while to outline the facts.

Mesmer's early career

Franz Anton Mesmer (1734–1815) was an almost exact contemporary of Hahnemann (1755–1843). He grew up on the shores of Lake Constance, on the border between Germany and Switzerland, in a Swabian town called Iznang. His father was gamekeeper to the Bishop of Constance and Mesmer was brought up as a Catholic; indeed, as a youth he contemplated entering the priesthood, but he soon reslised that he lacked a vocation. For a year he studied law, but in 1760 he became a medical student in Vienna, where he qualified MD and PhD in 1767 at the fairly advanced age of 32.

Mesmer was thus, like Hahnemann, well grounded in the science of his day, and he showed no leaning towards occultism or mysticism. It is therefore somewhat ironic that his name is now linked with these qualities.

His early career after qualifying was, in fact, conventional enough. He married a rich aristocratic widow, ten years older than himself, and thanks to his wife's connections soon established a prosperous practice in Vienna, where he met and became friendly with the young Mozart and his father. Not until the 1770s did he begin to move in the direction that was later to bring him such renown and notoriety.

A young girl called Franz Oesterlin, a relative of Frau Mesmer, became Mesmer's patient. She was suffering from symptoms that would now be regarded as psychological. In order to make herself more easily available for treatment she came to stay with the Mesmers, and as he studied her case Mesmer was led to formulate a remarkable theory.

Discovery of animal magnetism

Mesmer's doctoral thesis had been concerned with the influence of gravitation on human physiology. He had suggested that gravitation depends on a subtle universal fluid which he imagined to pervade the whole cosmos, including living organisms, and to set up "tides" in the bloodstream and nerves of human beings.

This thesis, which in later years he referred to as "The Influence of the Planets on the Human Body", sounds as if it should be concerned with astrology, but Mesmer intended it to be fully scientific. Ideas of this kind were acceptable scientific currency in the eighteenth century, and indeed Mesmer had lifted whole sections of his theory from the writings of the respected English physician Richard Mead.

Contemplating Franz's symptoms, he made the "obvious" connection. He now understood what was causing the ebb and flow of her attacks: nothing else than the gravitational tides he had described in his dissertation.

How to use this discovery to effect a cure? Why, by magnetism. Magnets were already in use by at least some doctors, though admittedly this was a contentious subject; and of course magnets, with their polar attraction and repulsion, could be plausibly supposed to act in the same general way as gravitation.

Mesmer's friend Maximilien Hell, professor of astronomy at the University, had a number of magnets made for him in the astronomy department, with different shapes according to the part of the body they were intended to treat. The effects were gratifying. As soon as the magnets were applied to Franz she had an immediate strong reaction followed by a dramatic improvement, and after further experiments Mesmer convinced himself that he had succeeded in controlling the ebb and flow of the universal gravitational fluid.

Almost immediately after this Mesmer quarrelled with Hell about who should have credit for the discovery. Hell claimed that it was the magnets themselves that had effected the cure, but Mesmer insisted that their only role was to channel the cosmic flow through the patient. It was in fact unnecessary to use magnets, he discovered; objects made of cloth or wood worked just as well.

The explanation, he concluded, was that he himself was touching them; he was an "animal magnet" who acted on objects and people in an analogous way to a mineral magnet acting on metal. Mesmer now tried to persuade the medical Establishment in Vienna of the validity of his discovery. In this he was unsuccessful, but Franz made a complete recovery and eventually married Mesmer's stepson. (Mozart, in a letter, records a meeting with this lady, now grown stout and the mother of three children.) Mesmer's fame increased, and so did his practice; in 1755 and 1776 he travelled in Swabia, Bavaria, Switzerland, and Hungary, treating the famous.

He was less successful in the case of Maria Theresa Paradies, a girl suffering from psychologically caused blindness since the age of three who was nevertheless a professional pianist. She had been treated with the conventional drastic methods of the time — bleeding, purging, blistering — and also with some experimental techniques, including the application of a tight plaster helmet and painful electrotherapy.

At first Mesmer was successful: Maria Theresa recovered her sight, at least temporarily. But the ophthalmologist who had failed to cure her was, not unnaturally, jealous of Mesmer, and claimed the cure was not genuine. Eventually, for reasons that are unclear, the patient's father reacted violently against Mesmer, finally appearing at his house, sword in hand and demanding that the treatment of his daughter be stopped.

Partly, at any rate, the explanation for the fiasco is that as the girl's sight improved her piano-playing deteriorated; she ceased to be so much of a public curiosity and was in danger of losing a pension that she received from the Empress. Perhaps, too, there were other causes connected with the Paradies' family life (sexual abuse?) which may have been responsible for the girl's initial hysterical blindness — the father's behaviour might suggest this. At any rate she relapsed; eventually she achieved a reasonably successful career as pianist and composer, but she never again recovered her sight.

Mesmer, meanwhile, was the centre of a scandal. Many people suspected him — almost certainly unjustly, since Mesmer seems to have had little interest in sex — of having had improper relations with Maria Theresa, and the hostility of the Viennese doctors increased. In 1778 Mesmer, by now informally separated from his wife, left Vienna for Paris.

Mesmer in France

Once established in Paris, Mesmer began a long series of feuds with the French medical Establishment. The Academy of Sciences, in spite of attending demonstrations, were unconvinced by the animal magnetism theory. Mesmer therefore approached the newly founded Royal Society of Medicine, which he hoped would be more amenable than the long-established Paris Faculty of Medicine.

His initial demonstration at his suite in the Place Vendôme was not well received. In 1778, therefore, he moved out of Paris and set up a clinic at a nearby town, Créteil, where he had more room to treat the large number of patients who flocked to him. Some received individual therapy, while the less seriously ill or the convalescent were treated in groups. For this purpose Mesmer invented the baquet, a large wooden tub containing bottles of magnetic metal, stone, glass and so forth. Mesmer had magnetised all these items himself, by touching or pointing at them. The baquet had iron rods projecting from it; the patients pressed these against the affected parts of their bodies, and they also held hands to allow the animal magnetism to flow through the group.

Many grateful patients wrote testimonials to the efficacy of the treatment, but the Royal Society was unimpressed and refused to attend the demonstrations. However, Mesmer was more successful with the Paris Faculty of Medicine, a prominent member of which, Charles Deslon, became a convinced believer in animal magnetism. He had himself magnetised, served as Mesmer's assistant, and eventually established his own clinic.

Having moved back again to Paris, Mesmer now accepted Deslon's suggestion that they should try to gain the endorsement of the Paris Faculty. Three prominent members of the Faculty agreed to watch Mesmer at work . They were shown a number of remarkable cures but remained obstinately unconvinced.

Mesmer now gave up hope of obtaining the Establishment's approval and concentrated on his clinical work. It is important to notice that he distinguished between what we would now call psychological and physical disorders, and refused to treat the physical. His patients ranged from the rich and aristocratic to the poor; everyone received an equal a mount of attention and those who could not afford to pay were treated free.

One feature of Mesmer's methods which attracted a good deal of unfavourable comment was the "Mesmeric crisis". Some patients, especially those suffering from more serious symptoms, experienced nervous trembling, nausea, occasionally delirium or convulsions. Mesmer regarded these as an inevitable accompaniment of the process of normalisation of the flow of animal magnetism, and he had special padded "crisis rooms" in which patients could throw themselves about without hurting themselves, while Mesmer or his assistants gave them individual attention. The depth of the crisis naturally varied from case to case, but Mesmer insisted that some degree of crisis, no matter how slight or transient, would always be found if it was looked for carefully enough.

Even more dramatic than the crisis, however, was the Mesmeric trance. Mesmer discovered this phenomenon only after he had been practising his method for some considerable time; the trance then became for him a method of inducing the crisis. Another of his followers, the Marquis de Puységur, discovered that it was possible to communicate with people in trance, getting them to answer questions, remember long-forgotten childhood events, and so on. The Marquis came to believe that it was possible to produce cures without a crisis, but Mesmer, constrained by the demands of his theory, did not agree.

Was Mesmer a hypnotist?

It is generally held that Mesmer was practising hypnotherapy, but it is probably more accurate to say that he was a shamanistic healer whose methods certainly included hypnotherapy but were not identical with it. Mesmer's conduct during therapy sessions was highly theatrical, being intended to augment the drama of the situation as much as possible. His clinic was meticulously furnished to maximise suggestion: the light was dim, everyone conversed in whispers, and music was used to alter the patients' mood according to what was required at each stage of the process. There were four baquets in the room, three for paying patients and the fourth for those being treated free.

Mesmer, as Master of Ceremonies, was elaborately dressed and carried a wand, which he pointed at patients or used to touch or stroke them. The patients gasped, twitched, went into trance, or experienced convulsions or catalepsy. Mesmer's assistants ministered to the more severely afflicted and if necessary brought them into one of the padded crisis rooms.

Disputes within Mesmerism

Although Mesmer made some influential converts, especially Deslon, he was eventually to break with almost all of them. He was autocratic and dictatorial (like Hahnemann) and would brook no opposition. A lawyer called Nicolas Bergasse became converted to Mesmerism and suggested to Mesmer the establishment of a private academy to propagate his ideas. The result was the grotesquely misnamed Societé de l'Harmonie.

The Society was secret. All the members had signed an undertaking that they would not pass on any part of Mesmer's teaching without his written permission, nor would they establish a clinic without such permission; they were allowed to treat individual patients only. It was this last condition that destroyed the Society within two years of its foundation in 1783.

The Society combined the roles of institute, medical school, and clinic. Students learnt the theory of Mesmerism and how to apply it in practice to patients. Schools were set up in Paris and also in several other cities in France, and thousands of pupils attended the courses. Bergasse took on much of the administration and became correspondingly powerful within the organisation.

Meanwhile Mesmer's erstwhile assistant Deslon had set up on his own account, and in 1784 he was investigated by a royal commission. The committee was convinced by his cures but denied, once again, the reality of animal magnetism. Another commission, set up by the Faculty of Medicine, reached the same conclusion. Mesmer objected that it was he, rather than Deslon, who should have been investigated, but there was nothing he could do about it.

Bergasse, Puységur, and other disciples of Mesmer now began to make public the knowledge of animal magnetism. Mesmer was furious, and the Society dissolved amid scenes of rancour and confusion. In any case the Revolution was coming and Mesmerism began to be overtaken by politics; Bergasse was later to adapt the doctrine of animal magnetism to support his views on revolutionary politics.

Mesmer kept aloof from politics. He travelled about in Europe for a number of years, though he was back in France from 1798 to 1802; he sued for his losses under the Revolution and was awarded enough to keep him in reasonable comfort for the rest of his life.

Declining years

He now recommenced his wanderings, and began to develop more outlandish ideas than he had entertained hitherto, starting to speculate on what we today would call paranormal phenomena and extrasensory perception. During the trance, he said, the mind comes into contact not only with other minds but also with the cosmos, and so in principle is capable of acquiring universal knowledge. In this way it is possible for seers and fortune-tellers to foretell the future. He published these ideas in a book in 1799, and as a result gained the reputation of an occultist. He died in Switzerland in 1815. He was in his eighty-first year; a gypsy in Paris had foretold long ago that he would die at this age, and he believed her, so he was prepared for the end when it came.

Assessment of Mesmer's ideas

In his own terms, Mesmer must be judged to have failed. His dominating ambition was to achieve scientific recognition for his theory of animal magnetism and this did not occur. His methods of treatment, however, were reinterpreted as suggestion and were rechristened hypnosis or hypnotherapy. In this form they were taken up by, among others, Jean Martin Charcot, Pierre Janet, and Sigmund Freud (although Freud later abandoned hypnosis). Although a faint aura of the disreputable has clung to hypnosis, there has always been a minority of doctors and psychiatrists who have valued and practised it.

Like Mesmer himself, some people have been attracted by the idea that hypnosis facilitates telepathy and clairvoyance. Numerous books, some by doctors, appeared in the mid-nineteenth century describing remarkable cases of thought transmission and other marvels during trance. It is interesting, however, that Mesmer's name is not mentioned at all in some of these books; the aura of charlatanry could not be dissipated . Matters were not helped by the development of hypnosis as a stage entertainment. There was also the fear that hypnotists might be able to manipulate their subjects for their own purposes; Svengali might be fictional, but could there not be real-life Svengalis?

Mesmer regarded his ideas as thoroughly scientific, although admittedly he did later flirt with the occult. In the nineteenth century hypnosis was part of the stock-in-trade of occultists such as Helena P. Blavatsky, the founder of Theosophy, and there is still a widespread belief that the hypnotic trance affords a way into hidden depths of the mind. And although the term "animal magnetism" is little used today, very similar ideas keep surfacing under other names: for example, Wilhelm Reich's "orgone energy".

Similarities

The sixth edition of Hahnemann's textbook *The Organon* contains a number of approving references to the then topical subject of Mesmerism. Hahnemann apparently used Mesmeric techniques himself, and he made a connection in his mind between the vital force which, he believed, brought about healing, and Mesmer's animal magnetism. Writings by homeopaths in the nineteenth century make references to Mesmerism, but later generations of homeopaths have made little of the connection, probably because of the reputation for charlatanry that later became attached to Mesmer's name.

The similarities between Mesmer and Hahnemann, both in career and in ideas, are surprisingly close. It's worth listing them.

- They were almost exact contemporaries.

- Both came from fairly humble backgrounds. Neither had very much to say about his childhood, which may have been because neither was particularly happy.

- Both qualified, rather late in life, as orthodox physicians and both adopted heterodox ideas that brought them into conflict with the medical Establishments of their day and came to dominate their lives and thought completely.

- Both spent a considerable time in Paris.

- Both had lawyers as prominent followers.

- Both started as scientists and then moved gradually towards more occult or mystical ideas.

- Both were characterised by feelings of injustice and persecution.

- Both were intolerant of any deviation on the part of their followers, with whom they became involved in acrimonious and destructive disputes, which led to the closure of establishments set up to propagate their ideas (Mesmer's Society of Harmony, the Homeopathic Hospital in Leipzig).

- Both insisted that cure must always be preceded by an aggravation or crisis, no matter how brief and slight.

- There are close resemblances between Hahnemann's vital force and Mesmer's animal magnetism. It is significant that some American homeopaths actually suggested the existence of a homeopathic force, which they called Hahnemannism by analogy with Galvanism and Mesmerism.

Hypnosis is used today, sometimes under other names, to treat certain disorders and even as an alternative to conventional anaesthesia for surgery, so it has achieved at least a degree of medical respectability. And modern methods of studying the brain have shown changes when people are hypnotised. It is too soon to know if homeopathy will follow a similar path towards semi-acceptance within conventional medicine. Much will depend on the outcome of scientific studies of the kind I discuss in Part II.

Part II

HOMEOPATHY TODAY

Chapter 11

Can We Prove Homeopathy?

Quite often we hear the claim that this or that piece of research has "proved homeopathy", or at least has provided support for it. Such statements often reveal a lack of knowledge about what homeopathy is. As I hope I've made clear, it is not a seamless unity but consists of a rather loose amalgam of ideas that can only be understood if they are viewed historically. This makes it rather difficult to do research on homeopathy itself, as opposed to research on individual homeopathic medicines.

Hahnemann's original revelation was the similia idea, which gave him a range of medicines to use and a way of choosing among them. Later he supplemented this with potency but there is no logical connection between the two; one could have homeopathy without potency or vice versa. It could be the case that homeopathic medicines work, at least in material doses, but ultra-molecular dilutions don't work, or alternatively highly dilute substances might have a demonstrable effect of some kind although there is no clinical benefit from using them. If the second possibility turns out to be the case, homeopaths will have stumbled on a curious phenomenon which has little or no relevance to treatment though it is of interest to physicists and chemists. It wouldn't be the first time that something of the kind has happened in science.

The two questions have to be studied separately, but a lot of the research that has been done has not been focused clearly on either of them. To complicate matters further, there are several different kinds of homeopathy. "Complex homeopathy", using mixtures of remedies, for example, has little in common with Kentian homeopathy apart from the name. Yet trials of "homeopathy" often lump them all together and sometimes even include treatments that could scarcely be called homeopathic at all. We shall see some examples of this shortly.

The similia principle itself, though supposedly central to homeopathy, is particularly difficult to test because the judgement of similarity is so subjective. When family members look at children they commonly see different resemblances. Aunt Mary will say that little Tommy looks just like his mother, Uncle Joe says he looks just like his father, and Cousin Anne says he looks like neither of the above but has a vague resemblance to Great-Aunt Emily. Clearly all these people are paying attention to different features and have different images in their minds. To a large extent, similarity lies in the eye of the beholder.

Experienced homeopaths often differ about which is the true similimum in any particular case, and a common objection to clinical trials of homeopathy which fail to show an effect is that the selection of remedies was not good enough — the true similimum wasn't found. One way of getting round this is to use isopathy, which allows the same remedy to be given to all the patients, but some say that this is not truly homeopathic. Another approach is to have a group of experienced homeopaths agree in advance on the appropriate remedies. Both methods have been used in clinical trials.

For all these reasons I think that the idea of "testing homeopathy" is pretty meaningless. Even if a particular homeopathic remedy is shown to work, this does not "prove homeopathy". At most, it shows that that particular remedy works. So we have to break homeopathy down into its components and test them separately. Research could be intended to investigate any of the following:

1. the potency effect — is it real? This can be tested on patients or it can be investigated in the laboratory.

2. homeopathic remedies — do they really have any effect on symptoms or disease?

3. provings — are they reliable and do they contribute new knowledge for homeopaths?

The first two are mainly directed at critics; the third is more for the "home market, to help homeopaths in their practice. In what follows I look at each of them separately. Please see the appendices for references to the studies I cite.

The potency effect

One of the problems that most kinds of alternative medicine face is that critics dismiss them on the grounds that they are self-evidently absurd. Homeopathy has always been especially vulnerable to such criticism, because its use of very small doses — so small, in many cases, that none of the original substance should be left at all — inevitably excites scepticism or downright incredulity and derision. Homeopaths are forced to resort to saying that information from the original substance is somehow preserved in the fluid used for dilution (normally an alcohol–water mixture); this is often referred to as the "memory of water" phenomenon. They can point to a certain amount of rather strange facts about the nature of water which might conceivably provide a physical basis for their claims, and various theories have been suggested. But these ideas are still very tentative and have not yet gained widespread acceptance by physicists. (See Appendix E.)

The Benveniste affair

For almost a hundred years homeopaths have attempted to demonstrate the existence of the "potency effect" scientifically. A recent example of this, which unfortunately ended in near-farce, occurred in 1988, when the French researcher Jacques Benveniste was "investigated" by the journal *Nature*. On 30th June 1988 the journal published an article by Benveniste and his colleagues at the Unit for Immunopharmacology and Allergy of INSERM at Clamart, in the outskirts of Paris. The article appeared to provide support for homeopathy.

When a certain type of human white blood cell, the polymorphonuclear basophil, is exposed to antibodies against IgE (the protein concerned in allergic reactions), certain changes occur. Histamine (the chemical that causes many of the clinical symptoms of allergy) is released from the cell, and the cell itself changes its appearance.

Benveniste and his team claimed that these changes could occur even though the liquid containing the anti-IgE antibodies was diluted to fantastically high levels, far beyond the point at which any molecules of the starting substance could be expected to be present. As Benveniste put it, perhaps rather over-dramatically, in an interview in *Le Monde*, it is as if one shook a car key in the Seine at the level of the Pont Neuf in Paris and then collected a few drops of water at Le Havre that would start that very car and not another.

Benveniste also found that in order to produce these effects it was not enough just to carry out a plain dilution; vigorous shaking, of the kind used in making homeopathic medicines, was required. Another interesting finding was that there were successive peaks and troughs in the effect as the dilution process was continued. (This feature has appeared repeatedly in homeopathic research as far back as the early 1900s, and presumably must mean something; it suggests a kind of "resonance" phenomenon.)

As an established scientist with a sound reputation, Benveniste was well aware of the storm of controversy that his paper was likely to provoke. However, he can hardly have been prepared for the scandal that broke over his head soon after his paper appeared. The editor of *Nature*, John Maddox, had accompanied publication of the paper with an editorial expressing considerable reservations:

> Benveniste's observations are startling not merely because they point to a novel phenomenon, but because they strike at the roots of two centuries of observation and rationalization of physical phenomena. The principle of restraint which *Nature* applies in its editorial is simply that, when an unexpected observation requires that a substantial part of our intellectual heritage should be thrown away, it is prudent to ask more carefully than usual whether the observation may be incorrect.

Benveniste was in full agreement that his results ought to corroborated by other scientists — indeed, this had already happened at five other institutions. But in a later television discussion he also said that there was no need to be quite so apocalyptic as Maddox had been in saying that two centuries of science would have to be thrown away. Benveniste's results, if correct, were certainly very interesting and important, but they were not quite as world-shaking as that. They were, he thought, in principle capable of being explained by the electromagnetic properties of water.

On 28th July *Nature* published what was in effect a retraction of its initial decision to endorse Benveniste's paper at least to the extent of agreeing to publish it. An investigative team, composed of John Maddox himself, James Randi (a professional magician and debunker of claims for the paranormal), and Walter W. Stewart (a specialist in errors and inconsistencies in the scientific literature and scientific fraud), spent five days at Benveniste's Unit at Clamart. Their report, entitled "'High dilution' experiments a delusion", was dismissive of his results. It concluded that:

> the care with which the experiments reported have been carried out did not match the extraordinary character of the claims made in the interpretation; the phenomena described are not reproducible, but there has been no serious investigation of the reason; the data lack errors of the magnitude that would be expected and which are unavoidable; no serious attempt has been made to eliminate systematic errors, including observer bias; the climate of the laboratory is inimical to an objective evaluation of the exceptional data.

In other words, Benveniste, in the view of the investigative team, had been guilty of extreme gullibility and self-deception.

Benveniste understandably reacted with great anger — not to the fact that an inquiry had been carried out, for he had been quite willing for this to be done — but to the way in which it had been conducted and to the implication that his team's honesty or scientific competence were questionable. "The only way definitively to establish conflicting results," he said, "is to reproduce them. It may be that we are all wrong in good faith. This is no crime, but science."

Several things occur to me about this sorry tale. One is that it seems extraordinary that a scientific journal like *Nature* did not conduct its investigations *before* publishing Benveniste's paper rather than afterwards. Another is that the composition of the team, which did not include anyone competent to assess Benveniste's work scientifically, must surely indicate the kind of conclusion it was expected to reach. A third is that it was naive of Benveniste not to anticipate this outcome when he was informed of the composition of the team; it was then that he should have objected.

Some of Benveniste's later work was, one has to say, considerably harder to fit into a scientific framework than the research I outlined above. It has been described by Michael Schiff, a physicist at CNRS, the French National Centre for Scientific Research who worked closely with Benveniste. In his book *The Memory of Water: Homeopathy and the Battle of Ideas in the New Science* Schiff claims that Benveniste demonstrated that it is is possible to transmit information about biological preparations electronically, via a "black box".

In outline, the setup was follows. A tube containing the test material, for example the white of an egg, was placed in a coil. This was connected to the black box, which in turn was connected to another coil enclosing a second tube. Water from this second tube was assessed for biological activity in what is called a Langendorff apparatus. This contains the heart of a freshly killed guinea-pig or rat that has been immunised against egg albumin (in this example); measurement of activity then depends on estimating the rate of flow of the test fluid through the coronary vessels of the heart. (This is a standard technique in physiology research.)

Schiff maintains that it is not important to know what the black box actually contained or how it worked and no details are given. In fact, it seems that more than one kind of transmission apparatus was used. The original machine was provided in June 1988 by a homeopathic doctor called Attias; this was just before the visit by the *Nature* delegation. Later, Benveniste had his own machine built; all we are told is that it was "essentially a low frequency high gain amplifier". On the basis of numerous double-blind experiments, Benveniste (and Schiff) became convinced that it is indeed possible to transmit biological information electronically in the manner outlined above.

Having summarised these studies, Schiff goes on to discuss at some length why it is that the scientific community at large has not accepted the validity of the work in question. His argument, in brief, is the fairly well-worn one that science is a closed shop and rejects any new ideas that do not fit into its current world picture. Dismissal of Benveniste's claims about the memory of water are, he says, merely one aspect of a wider refusal to consider the possibility that contemporary science could be wrong. This is essentially a conspiracy theory.

How well do Schiff's arguments stand up? Certainly it isn't difficult to think of numerous instances from the history of science which support the thesis. One of the most recent and striking of these was the rejection for many years of Wegener's theory of continental drift; it has now become the cornerstone of geology. It is also easy to think of instances of claims for dramatic discoveries that have *not* been substantiated. Which of these categories Benveniste's work will finally fall into is still uncertain, but I cannot think that Schiff's book will do much to hasten its acceptance by orthodoxy.

Research on the "memory of water" goes on and positive results are still being reported. An international organisation called GIRI, founded in 1987 by Professor Madeleine Bastide of the University of Montpellier, exists to provide a forum for such work, which continues in 20 countries. A recent example comes from Madeleine Ennis and her colleagues in Belfast, who found that high (ultra-molecular) dilutions of histamine had effects on basophils which were similar though not identical to those reported by Benveniste. Ennis says she is a sceptic about homeopathy (Appendix E).

Provings

Homeopathy is often said to be based on provings, and while this is by no means wholly the case, since much of it derives from "clinical" symptoms, it is true that it started in this way. Provings are still carried out today, both to try to verify Hahnemann's findings and to test new substances for possible use as medicines.

Hahnemann, as we know, tested his medicines on supposedly healthy people; this was mainly how he built up his knowledge of the medicines. In his day the power of suggestion was not appreciated fully, and this has led many later critics to question the validity of the results he and his associates obtained. Of course, some of the homeopathic medicines are made from substances such as arsenic, phosphorus, and lead that are well-known poisons, and there is no doubting the symptoms that the nineteenth century provers report from these. Some of them, with commendable heroism, took considerable quantities of toxic substances for weeks or even months to observe the effects, and these reports make fascinating reading today. (See Chapter 4 for details.)

Not all the medicines used in homeopathy are obviously toxic; common salt, for example, would hardly be expected to have dramatic effects, and the same is true of some of the herbal substances. What some modern provings of these relatively harmless substances has shown, however, is the extraordinary power of self-suggestion. Lest it appear that I am being unduly sceptical here, I will describe briefly what happened in a modern proving, carried out in 1978 (Appendix C). The aim was to apply modern statistical methods to the analysis of provings and the substance chosen for testing was pulsatilla. This is very widely used in homeopathy and was extensively proved by Hahnemann and others.

The proving was carried out with a 3x potency — that is, with a low (material) dilution. This was chosen instead of the undiluted tincture because it was the strongest preparation that would not have an identifiable taste or appearance. The proving was carried out on volunteers in the north-west of England; most were members of a large philosophical society and were interested in homeopathy, though their actual experience of it varied greatly. The fact that most of the provers knew one another was a drawback, but the same was true of most of the nineteenth-century provings.

The trial was planned to last three months, with provers taking one tablet twice daily and recording their symptoms in a diary. During the first month all the provers received a dummy tablet (placebo); they did not know this although Dr Clover, who was conducting the trial, did. In the second month half the provers received pulsatilla and half received dummy tablets, while in the third month those who had received dummy tablets now received pulsatilla and vice versa. In the second and third months neither the provers nor Dr Clover knew which provers were receiving pulsatilla, and indeed at this time the provers did not even know the name of the remedy that had been chosen for the trial.

The results were remarkable. Thirty of the fifty-two participants returned their diaries filled in to some extent although only 18 completed the whole 3-month trial. There was no evidence that pulsatilla had produced any more symptoms than had the dummy tablets. What was very striking, however, was the fact that much the largest number of symptoms occurred during the first month — that is, at the time when *all* the volunteers were taking dummy tablets. Some, indeed, withdrew from the trial because of the severity of their symptoms. The incidence of symptoms declined progressively over the whole 3-month period regardless of whether provers were taking pulsatilla or dummy tablets.

This trial does not necessarily show that pulsatilla 3x is incapable of producing symptoms but in this instance any symptoms it did produce were totally swamped by the enormous number of placebo (or "nocebo") symptoms. This will not come as any great surprise to orthodox doctors, who are by now well aware of the importance of the placebo effect, but it does reinforce the point that the older proving literature has to be viewed with a lot of caution. True, the more critical writers of the time, such as Richard Hughes and Robert Dudgeon, recognised this and allowed for it as best they could, but in many cases it's almost impossible to assess the reliability of the reports. This applies particularly to the provings of the relatively inert substances, among which are some of the most widely used homeopathic medicines.

Another example of the same kind of effect was reported in a letter to *The British Homeopathic Journal* (Appendix C). Dr H. Walach, from the University of Freiburg, was giving a lecture on provings, in which he was describing his own experience with a homeopathic medicine derived from the rattlesnake (lachesis). During the lecture he invited the audience to take part in an experiment; he handed round two bottles containing pillules; the bottles were labelled simply 1 and 2. Nine people took the pillules.

Ten minutes before the end of the lecture Dr Walach asked what effects people had had from the medicine. Of the 9 who had taken the pillules, 4 had had definite symptoms — two with preparation 1 and two with preparation 2. Some were quite striking: one person had felt the whole left side go to sleep and had experienced a choking sensation. These are the effects that would be expected from lachesis, according to the homeopathic literature (lachesis is a "left-sided" remedy). The audience was then asked to say which bottle they thought had contained the real medicine; their assessments were equally divided. And the answer? You've probably guessed it: neither — both were placebo.

Aggravations

From an early stage in his description of homeopathy Hahnemann taught that the correctly chosen remedy would always produce a temporary worsening of the patient's symptoms. Subsequent generations of of homeopaths took up this idea and frequently described seeing such aggravations in their patients. Some were sceptical, however: Robert Dudgeon thought that aggravations were much less common than Hahnemann supposed.

In trying to decide about the reality of aggravations we have to remember that when Hahnemann first started to treat patients homeopathically he was using material, though admittedly small, doses of the medicines. It is therefore quite possible that these did sometimes produce symptoms and Hahnemann could have called them aggravations. Later he changed to using high potencies (ultramolecular preparations) that would not be expected to contain any of the original molecules of the remedy and it is unlikely that they would have caused symptoms directly.

I was interested in aggravations and looked for them for more than twenty years, finding only one possible instance in all that time. I think this was because I never told patients they were likely to experience an aggravation. Other prescribers did see them quite frequently but they regularly warned their patients about the possibility. I think myself that apparent aggravations are due to suggestion, which causes the "nocebo" effect (opposite of the placebo effect). In one case a patient reported experiencing a severe psychotic episode after taking a placebo dose consisting only of sugar.

Assessing clinical effectiveness

Interesting and important though basic scientific research and provings may be for homeopaths, what really matters is to show that the treatment is actually useful in practice. That is, it must be subjected to scientific evaluation in a controlled setting. The standard way of doing this is in a "randomised controlled trial" (RCT). In outline, such trials are done as follows.

A comparison is made using a number of patients, who are divided randomly (this is important) into groups who receive either the active treatment, a placebo, or a different treatment. Whenever possible the trial is "double blind", which means neither the patients nor the doctors know who is receiving the "real" treatment. The results are then analysed statistically. The patients are matched as carefully as possible for age, severity and duration of symptoms, and other factors so that all the groups are comparable. The detailed design of such studies is a complicated business and there is often argument about the interpretation of particular trials, but it remains the case that clinical research has to be based on methods of this kind nowadays if it is to be taken seriously.

There has been a lot of discussion about the feasibility of carrying out RCTs in homeopathy. It is said that because the treatments are individualised it isn't possible to compare groups of patients in the standard way. Attempts have been made to find other trial designs that would be fairer to homeopathy, but these carry less weight with non-homeopaths. It's therefore been necessary for homeopaths to go along with randomised controlled trials, however reluctantly, and quite a few have now been done, with varying results. In fact, the sheer amount of research that is available makes it difficult to evaluate, especially as much of it is of poor quality and possibly subject to bias.

A standard way of trying to draw conclusions from a body of research like this is to carry out what is called a *meta-analysis*. The idea is to look at a large number of trials and to rank them according to preset criteria to try to assess their reliability. A scoring system is applied, with each paper being given marks for such things as adequacy of description of the methods used, presence or absence of "blinding", number of patients studied, and so forth. Not everyone is fully persuaded of the value of meta-analyses but they are widely used in modern medicine. Only in this way is it possible to sift the vast amount of research that now is carried out throughout the world.

In what follows I shall look at both meta-analyses and clinical trials in homeopathy, to try to give what I hope is a balanced idea of the issues. But be warned: this is a large subject and, moreover, one that is constantly changing, so I can't hope to cover every aspect of it.

A. Meta-analyses

A number of meta-analyses of homeopathy have been published (Appendix B). The first, by Kleijnen and colleagues, appeared in the *British Medical Journal* in 1991 and was fairly encouraging, although it did find that the better studies were also the ones least likely to show a positive effect. (This is a frequent comment in meta-analyses of trials in all kinds of unconventional medicine.) But the study which is most often quoted by homeopaths was by Linde and colleagues and was published in *The Lancet* in 1997. Its rather cautious conclusion was that the results "are not compatible with the hypothesis that the clinical effects of homeopathy are completely due to placebo". Interpreting this requires careful navigating through the double negatives but it does imply that homeopathy may have real effects over and above the placebo response. The review was not wholly positive for homeopathy, however. It said that there was some evidence of publication bias in the available literature and that it was not possible to say that homeopathy was effective for any single clinical condition. As usual, further research was called for.

Some more recent meta-analyses have been less encouraging. One by Cucherat and colleagues, published in *The European Journal of Clinical Pharmacology* in 2000, found "some evidence that homeopathic treatments are more effective than placebo, but the strength of this evidence is low because of poor trial quality". In 2002 Professor Edzard Ernst, head of the Department of Complementary Medicine at the University of Exeter, carried out a "systematic review of systematic reviews" of homeopathy and concluded that "the best clinical evidence to date does not warrant positive recommendations for its use in clinical practice". And *The Lancet* has now backtracked from its earlier partial endorsement of homeopathy and is very critical of it. In 2005 it published a meta-analysis by Shang and colleagues which reported: "Weak evidence for specific effect of homeopathic remedies, but strong evidence for specific effects of conventional interventions. This finding is compatible with the notion that the clinical effects of homeopathy are placebo effects." An editorial in the same journal was dismissive of homeopathy.

As the Faculty website acknowledges: "It is fair to say that the meta-analyses ... of homeopathy are pretty inconclusive ... at present the current meta-analyses do not provide sufficient information on which to base an opinion about homeopathy in general." They therefore choose to present "some good single randomised controlled trials" as evidence for homeopathy. Let's see what they tell us.

B. Clinical trials

In 2007 the Faculty of Homeopathy website listed five conditions — asthma, influenza, glue ear, acute low back pain, and upper respiratory tract infection — to illustrate the effectiveness of homeopathy. I shall consider each of them briefly and also summarise some other trials which I think are of interest. (Please see Appendix D for a link to the Faculty website and references to the papers I cite here.)

1. Asthma

This randomised double-blind study by Reilly and colleagues in 1994 looked at 28 asthmatic patients with house dust mite allergy. They were divided into two groups to receive either homeopathic immunotherapy (isopathy) for their principal allergen, chosen by skin testing, or a placebo. Conventional treatment was continued as well. A significant (P=0.003) difference in favour of homeopathy was found for symptoms (visual analogue score), There were similar trends for bronchial reactivity and lung function. Improvement began after a week and persisted for up to eight weeks.

One of the criticisms of homeopathic research is that trials are seldom repeated to see if the initial findings were correct. There is an exception in this case, however, because in 2002 a study by Lewith and colleagues (not cited on the website) looked at the use of ultra-molecular homeopathic immunotherapy to treat asthmatic people with house dust mite allergy. It failed to confirm the earlier findings: in this case homeopathic remedies were no better than placebo in the treatment of patients who were allergic to the mites. The new study was a lot larger than the earlier one (over 200 patients took part) and it used a wider range of outcome measures. One curious finding was that in the third week the patients receiving homeopathic treatment felt their asthma to be worse and were more depressed than those receiving placebo.

In 2003 another asthma trial, by White and colleagues, also reached negative results. The effects of individualised homeopathic remedies in addition to conventional treatment were compared with placebo in 96 children with mild to moderate asthma. No significant differences were found between homeopathy and placebo.

2. Influenza and influenza-like symptoms

This report by Ferley and colleagues describes the use of Oscillococcinum to treat patients in an influenza outbreak. Oscillococcinum is a curious remedy derived from the liver and heart of ducks and its status as a homeopathic medicine is dubious to say the least. Its origins are unusual even for homeopathy, which has plenty of unusual remedies, such as dog's milk and the south pole of a magnet. Oscillococcinum was contributed to the homeopathic materia medica by a French doctor, Joseph Roy, in the pandemic influenza outbreak in 1917. Roy thought he had found organisms in the blood of victims which he called oscillococci. Subsequently he identified the same germ in patients suffering from numerous other illnesses including cancer. No one else has confirmed this discovery and, whatever Roy may have seen, it was not the cause of the diseases in question — least of all cancer. But Roy believed he could use his finding to cure cancer homeopathically.

All he needed to do was to locate a source of oscillococci and prepare them homeopathically as a nosode. This wasn't difficult, since they were supposedly abundant in nature, but for some reason he decided to use the liver and heart of ducks. This is still the source used today. The material is allowed to macerate for 40 days, after which it is potentised and used as a remedy — not for cancer but for influenza.

I think that a remedy as strange as this can be called homeopathic only as a matter of courtesy. It was not arrived at through proving and it is not even a nosode in the usual sense. Not for the first time, "homeopathy" seems to be understood very loosely by some nowadays. One almost gets the feeling that a "homeopathic" remedy is anything that a self-styled homeopath chooses to use. I cannot think that Oscillococcinum deserves inclusion in a trial of homeopathy.

3. Glue ear

Harrison and colleagues looked at homeopathy as a treatment for glue ear in 33 children. Although there was a trend in favour of homeopathy compared with placebo the difference did not reach conventional levels of statistical significance and the authors calculate that 270 children would be needed to show a difference of this magnitude. Nothing definite can be concluded from this study, therefore.

4. Acute low back pain

Stam and colleagues compared a "homeopathic gel" containing "Spiroflor" with another cream containing capsicum in the treatment of patients with acute low back pain. Both were equally effective although Spiroflor produced fewer adverse effects.

Spiroflor is a Dutch compound containing a mixture of homeopathic remedies (Symphytum, Rhus toxicodendron and Ledum). Traditionally, homeopathic medicines are given singly and orally and I think it is doubtful whether mixed external applications can be considered homeopathic treatment. Almost all homeopathic medicines today are potentised, often to a high degree, but the external applications contain either tinctures or 1x potencies (the lowest possible dilution). I should say that Spiroflor and similar compounds should be classed as herbal medicine rather than as homeopathy.

In any case, Spiroflor was no better than capsicum in this trial and it is entirely possible that both groups had a placebo response since there was no placebo group for comparison.

5. Upper respiratory tract infections in children

In this study of 170 children de Lange de Klerk and colleagues found that individually prescribed homeopathic remedies added little to careful counselling in reducing the "daily burden" of symptoms or need for antibiotics or surgery, but they may have "helped recovery from illness."

In summary, two of the five studies cited by the Faculty don't seem to have much to do with homeopathy and the remaining three showed only modest effects at best. The most positive outcome was for asthma but two subsequent trials with larger numbers of patients failed to replicate this result.

A long-term study

RCTs are usually short-term. Homeopaths have objected that this doesn't give their treatment a fair chance, because homeopathic remedies are often said to take a long time to act. This idea was tested in an interesting trial carried out by a team of six homeopathic doctors in Germany led by Dr Walach. The study was done on patients suffering from chronic headaches, who were recruited by a publicity campaign. In the first phase, which was double-blind randomised-controlled and lasted 12 weeks, no difference was found between homeopathy and placebo. To answer the criticism that this was too short a time for homeopathy to have an effect the doctors went on observing the patients for a further year.

The main finding in this second phase was that most of the clinically important changes had happened by the end of the 12 weeks' double-blind study, so more prolonged treatment didn't make much difference. Although many of the patients who responded had carried on with homeopathic treatment the results were only slightly better at one year. The authors think that 30 per cent of the patients could be classed as having responded. (This is the figure usually quoted as the average placebo response, although there is some argument about this — see below.)

A curious finding was that those patients who stopped all treatment, homeopathic or other, did best. Walach believes that this may be because homeopathy prevents patients from using harmful treatments and "allows the system to rebalance itself". He suggests that this may be key to understanding homeopathy in general — an intriguing thought. The failure of long-term treatment to improve the results significantly is, as the authors remark, "a quite provocative finding which should be carefully considered by the homeopathic community".

The treatment paradox

From the clinical research I have summarised here it is clear that we have a puzzle on our hands. There is a big gap between what occurs in practice, where both patients and practitioners are often enthusiastic about the results, and what can be demonstrated by research. How can we explain this apparent paradox?

Critics list a number of possible reasons why homeopathy might appear to be effective even if the remedies really do nothing. Here are some of them.

1. Spontaneous remission or recovery

Many disorders fluctuate in severity or clear up spontaneously. If this happens soon after taking a remedy it is psychologically difficult not to attribute the improvement to the remedy. There is a statistical phenomenon called "regression to the mean" which is relevant here. The more severe a patient's symptoms at the outset, the more likely it is that he or she will experience a reduction in their severity later, simply by the law of averages.

2. Concomitant orthodox treatment

Patients often continue to take orthodox treatment in addition to homeopathic medication, but it is usually the homeopathic treatment that gets most of the credit for any improvement that may occur. This is particularly the case for cancer but much the same can happen in other diseases.

3. Cessation of harmful orthodox treatment

As suggested by Walach and colleagues in their headache study, it can sometimes be beneficial to stop conventional treatment because it may be causing unwanted effects (side effects) or even worsening the condition. Many people with frequent or continuous headaches are getting them *because of* the analgesic medication they are taking regularly rather than in spite of it, and stopping the medication may cure their headaches. Homeopaths, particularly non-medical homeopaths, often advise their patients to stop their conventional treatment. This can of course be dangerous, if the treatment in question is essential for their health, but sometimes it is beneficial.

4. Associated life-style changes

Many, probably most, homeopaths do not confine themselves to giving their patients remedies but also advise them about their life-style. In many case the advice they give is similar to that which a conventional doctor would give (stop smoking, eat a more balanced diet, take regular exercise, get enough sleep), but patients are usually more prepared to take such advice when it comes from a homeopath. These changes may well be beneficial, regardless of any effect there may or may not be from the medicines that are prescribed.

5. Desire to please the therapist

Many patients, particularly patients receiving unconventional treatment, seem to wish to please the therapist by exaggerating the amount of benefit they are receiving.

I had a vivid experience of this once, when I was treating a man who had a neurological problem as a result of herpes zoster (shingles). He lived over a hundred miles away and was coming to the hospital for treatment (with acupuncture) every two weeks. He kept telling me he was improving so I went on bringing him back. Eventually, after several visits, he came accompanied by his wife because he had had a slight stroke. I happened to see the wife alone while he was out of the room and I asked her how he was. She said he was no better. When he came in I asked him about this and he admitted that it was true. He was such a pleasant man that he had felt embarrassed to tell me that I was doing him no good, so he had continued to suffer the expense and discomfort of a long train journey rather than disappoint me.

6. Placebo response

This is the principal reason advanced by critics to explain apparent homeopathic effects. It needs to be looked at carefully, because the question is not as simple as it sometimes appears to be. (See Appendix G for references.)

There are numerous myths about placebos. It is widely believed, for example, that placebo responders are "more suggestible" than other people. This is not true; anyone can be a placebo responder. It is often said that placebo effects are short-lasting (less than 6 weeks), but in some cases they can last for years. Although it is often said that almost any medical disorder can respond to a placebo, this may not be true; at any rate, pain, inflammation, and anxiety and depression seem to do best.

Many sources tell us that the response rate to placebos is about 30 per cent. This figure comes from a paper by Henry Beecher published in 1955. It is widely quoted, because Beecher is highly respected as one of the leading advocates of randomised controlled trials after the Second World War, but its reliability is questionable; studies have shown placebo effects to vary between 0 and 100 per cent in different trials.

Even the widely accepted equation between placebo and belief may not be entirely correct. A fascinating study was done by L.C. Park and L. Covi in 1965 at Johns Hopkins University, in which 15 patients suffering from anxiety were told they were having sugar pills yet 13 of them improved. Perhaps the patients disbelieved the doctors who said they were giving them a placebo, but in any case this trial shows how difficult it can be to draw conclusions from studies of this kind.

Defenders of homeopathy often point to the fact that it appears to work in animals and babies as evidence that its effects cannot be due to placebo. This is superficially persuasive, but of course the owners of the animals and the parents of the babies generally do know that treatment is being given and it is likely that their attitudes and expectations communicate themselves to the patients. Some of the non-specific effects already mentioned that are unconnected with the placebo effect may also operate in such cases. (I return to the placebo question and its relevance to homeopathy in Chapter 12.)

So where do we stand?

It is wrong to say, as some critics do, that there is no objective evidence for homeopathy. There is, but most of it is of rather poor quality. Even at its best there is evidence for only a small effect, and when an effect is as small as this it may not be there at all. It is also disturbing that the better the quality of a trial the less likely it is to show a positive effect.

Another worrying feature, from the enthusiasts' point of view, is that all the different forms of homeopathy — complex, "classical" (i.e. Kentian), or isopathic — seem to have about the same success rates. One would expect to see differences if their respective advocates were to be believed.

Some recent trials of "homeopathy" have used creams applied externally to the skin or even intra-articular injections, but it is doubtful if these could be described as trials of homeopathy at all (Hahnemann would have been horrified). The boundary between homeopathy and herbalism (phytotherapy) was always indistinct but now it seems to be threatening to disappear altogether. And some of the trials, such as those of Oscillococcinum, use remedies whose provenance is so obscure that I don't see how they can be called homeopathic.

I conclude that there are no firm answers to questions about the efficacy of homeopathy to be found in the research that has been done up to now. Homeopathy has not been proved to work but neither has it been conclusively disproved — it is of course notoriously difficult to prove a negative. Critics and enthusiasts alike are still feel entitled to make their own minds up.

The study of asthma by Lewith and colleagues which I outlined a short while back appeared in the *British Medical Journal*, where it was accompanied by an editorial by Gene Feder and Tessa Katz, who are both sympathetic to homeopathy. They admit that "The aggregated effect size of homeopathic treatments, when possible publication bias is taken into account or only high quality trials are included, is modest", but they still think that the refusal of critics to take homeopathy seriously is a consequence of their prior beliefs; such critics will not be convinced no matter what evidence is produced. They therefore suggest that it is no longer a research priority to carry out trials designed to test whether homeopathy is more effective than placebo. Instead, resources should be directed towards outcome studies to assess "the effect and cost effectiveness of homeopathic treatment".

For understandable reasons this is a popular view among homeopaths. Outcome studies regularly show patients to be satisfied with homeopathic treatment so "effect" in that sense is easy to demonstrate. Homeopathic remedies are relatively cheap, so on that score homeopathy comes out ahead of conventional medicine. The main cost disadvantage is time — homeopathic consultations are generally lengthy — but it is often unnecessary for the practitioner to have a conventional medical qualification, which can help to reduce the cost. Kentian homeopathy, in particular, is suitable for use by nurse practitioners or by homeopaths without a conventional medical background — in fact, there is a view that such people are better suited to homeopathy than are most conventionally trained doctors. So it should not be difficult to demonstrate that homeopathy is usually cost-effective.

I can go along with this utilitarian view up to a point, although it is vulnerable to the objection that, when money is tight, unconventional medicine is perhaps a luxury that the National Health Service can no longer afford. But I can't agree that it is no longer necessary to carry out research to find out if homeopathic remedies have objective effects. I do want to know. Willing suspension of disbelief can take you only so far.

I have to say that, from my own experience of it, the effectiveness of homeopathy — whatever it may depend on — is smaller than is often claimed. An enthusiast could of course object that this was my fault rather than the fault of homeopathy, and obviously that is possible. Perhaps I'm just no good at it. But I don't base my opinion simply on the results of my own efforts. During my training I was privileged to "sit in" for over a year at the clinic of an acknowledged homeopathic expert. During that time I saw a great deal of human warmth and personal attention being given to the patients, who were deeply grateful for it and felt that they were benefitting as a result, but I cannot honestly say that in any of these cases was there convincing objective evidence of improvement.

My own assessment of homeopathy is that, while it is impossible to say categorically that *all* the remedies are without objective effect, any effect there may be is small and unimportant. The great majority, at least, of the improvement that patients experience is due to non-specific causes. I look at this in more detail in my final chapter.

Chapter 12

Assessing Homeopathy

Homeopathy as placebo

There is no doubt that homeopathy is likely to be a strong placebo. If we consider the "standard" homeopathic consultation, by which I mean the Kentian version, it is undoubtedly well suited to maximise the placebo effect, for a number of reasons.

First, it takes a long time; most homeopaths like to allow at least 45 minutes for a first consultation and many prefer an hour or more. Prolonged personal attention is one of the strongest elements in producing the placebo response.

Second, patients feel that they are being treated "as an individual". They are asked a lot of questions about their lives and their likes and dislikes in food, weather, and so on, much of which has no obvious connection with the problem that has led to the consultation. Then the homeopath will quite probably refer to an impressively large and imposing source of information to help with choosing the right remedy. In the past this would probably have been Kent's *Repertory*, a large thick book like a dictionary. Today it may well be a computer, for programs now exist which allow the homeopath to refer not only to Kent's material but also to several other compilations of homeopathic lore. Unkind critics have seen a resemblance here to consulting the *I Ching* or casting an astrological horoscope, both of which, like homeopathy, are procedures centred on the individual who is their recipient.

Whatever adds to the ritual serves to enhance its efficacy. At one time I used to practise homeopathy privately in the rooms of a colleague who had a lot of remedies that were about fifty years old. They had been prepared long ago by a homeopathic doctor who had made them by hand, and they had been preserved by a process known as grafting. The medicines consisted of lactose powders, contained in bottles with handwritten labels and neatly stacked in rows on the shelves of cabinets.

The "grafting" consisted essentially in adding fresh lactose to the almost empty bottle, perhaps with a little alcohol–water mixture, and shaking it for a short time. This procedure was supposed to transmit the energy of the medicine to the added lactose — to potentise it, in fact.

This was quite a common procedure in earlier times but it would generally be frowned on today; modern homeopathic medicines are made with strict "quality control" to ensure their effectiveness. The starting "mother tincture" is assessed for purity and the process of alternate dilution and succussion is carried out according to strict rules. The whole manufacturing sequence is carefully regulated to ensure that the medicines are made in the correct way.

Theoretically, medicines prepared with all these safeguards presumably should be better. Our experience was, however, that patients nearly always seemed to respond better to the grafted medicines which we prepared ourselves rather than to those which we had sent from the homeopathic pharmacies.

We would carefully tip out some of the granules from the bottle onto a little square of clean white paper, which we would fold into a packet. Typically there would be several of these packets, perhaps five or seven, which were numbered to be taken daily in sequence. Patients would watch us making these preparations and then carry the medicines away reverently to take later at home.

We were ourselves surprised at the consistently better results we obtained with our old medicines, but we weren't alone in this; other homeopathic doctors who used the same procedures, though with different stocks of medicines, also found that method to be better. Was this because the hand-made medicines of yesteryear really had some magic ingredient that their modern marchine-succussed counterparts lack? Possibly. In retrospect, however, I think it more probable that the ritual of preparation witnessed by the patient was in itself impressively therapeutic and that the superior efficacy of our medicines was due to this.

So are we to conclude that homeopathy is simply a powerful placebo? Probably, yes, but there is nothing wrong with being a placebo responder! Wendy Kaminer, whom I quoted in the Introduction, was right about this. Anyone may respond to placebos, and if you do, it doesn't mean you are gullible or suggestible. Nor is the placebo response in any sense unreal. If you believe, as I do, that the mind depends on the brain, it follows that there must be brain changes when someone responds in this way. And we are now beginning to understand how this occurs, thanks to the extraordinary methods of visualising brain function that are becoming available. The brain areas responsible, especially those often lablelled limbic, are being identified (see for example Carlo A. Porro and also A.K.P. Jones and colleagues — Appendix G). So the placebo response is in no sense unreal or imaginary.

There are ethical problems attached to the use of placebos, but they are still prescribed quite regularly by many conventional health practitioners. A survey in Israel in 2004 found that, of 58 physicians and 31 head nurses questioned, 53 (60 per cent) used placebos at least occasionally (Appendix G). Among the users, 33 (62 per cent) did so at least once a month, 36 (68 per cent) told patients they were receiving actual medication, 15 (28 per cent) found placebos were useful in diagnosis, and 48 out of 51 (98 per cent) found placebos to be generally or occasionally effective. There is no reason to suppose that the use of placebos by health practitioners is confined to Israel, and for those who favour the practice homeopathy would seem to be a useful resource, since many of their patients believe in it even if they themselves don't. One very experienced homeopathic doctor, with a strong research background, has told me that he would continue to use homeopathy even if research were to show that the medicines had no objective effect.

But if homeopathy is a placebo, it is a complex one. There is a good deal more to it than just giving someone a remedy. I should say there are important similarities between homeopathy and psychotherapy — counselling. Both allow patients an opportunity to talk about their problems to an attentive and sympathetic listener *in a structured environment.* I think it's worth exploring the connections between homeopathy and psychotherapy a little further.

Homeopathy as psychotherapy

Psychotherapy is defined as "the talking cure", and judged on that basis, homeopathy includes a lot of psychotherapy. This is true whether or not the homeopath recognises that she is using psychotherapy. Many homeopaths would agree that there is an element of psychotherapy in the consultation though they would not accept that that is the main part of it. But homeopaths generally pride themselves, often with justification, on being people with good powers of intuition and empathy; indeed, unless they have these abilities they will not succeed in their profession. This also means that they are good psychotherapists.

Psychotherapy today uses many different theories but it originated with Freud and psychoanalysis. The psychiatrist Anthony Storr was sceptical about much psychoanalytic theory but nevertheless thought that psychoanalysis could have beneficial effects on patients. Dylan Evans comes to the same conclusion in his book *Placebo: The Belief Effect.* While working as a psychotherapist in the 1990s he obtained an impressive cure in a patient suffering from panic attacks by providing a psychoanalytic explanation for them. and he took this to be confirmatory evidence for the truth of the theory. But now he questions the cause-and-effect relationship that seemed convincing to him at the time. Perhaps the man's recovery was mere coincidence, or perhaps it was due to his belief that Evans's explanation was correct even though it wasn't.

I have had similar experiences myself. In the 1980s I quite often diagnosed patients as suffering from the hyperventilation syndrome, which was a popular label at the time, particularly in complementary medicine. The physiological explanation for it was quite plausible. Patients were supposed to have got into the habit of breathing a little too deeply, perhaps initially owing to mild anxiety. As a result they exhaled too much carbon dioxide with a consequent shift in blood pH towards alkalinity.

This allegedly produced numerous symptoms, including panic attacks, cramps, fast heart rate, altered sensations such as pins and needles, and even convulsions. To diagnose it we used the hyperventilation provocation test, which consisted in asking the patient to hyperventilate for two minutes; reproduction of the symptoms was taken to be a positive result. The accepted treatment of the acute attack consisted in advising the patients to rebreathe their own air using a paper bag; for a more lasting cure they were taught breathing exercises to favour abdominal diaphragmatic breathing over thoracic breathing.

The results were sometimes astonishing. One woman I saw used to have attacks in which she fell to the ground in the street and was unable to move for up to two hours. By doing no more than explain the supposed cause to her I was able to bring about a complete cure: no further attacks. I could not have wished for a more dramatic demonstration of the truth of the theory. But alas, it seems I was wrong; the evidence now indicates that the hyperventilation syndrome is probably a chimera. (See Appendix F.) In one well-executed study the hyperventilation provocation test showed no correlation between blood carbon dioxide levels and patients' symptoms and it is not alone in reaching this negative conclusion; several others have yielded the same result.

Another example occurred soon after I started using homeopathy. A woman was taking a number of psychotropic drugs which had given her face a mask-like appearance. She had already attempted suicide by jumping out of a first-floor window and breaking both legs. I gave her a single high-potency dose of thuja. She returned a couple of weeks later, looking and sounding totally transformed, and having stopped all her conventional medication, though I had not told her to do this. She remained well and happy for a year subsequently.

Is it conceivable that a single dose of thuja could work such a remarkable transformation? Some homeopaths might say so, but I think it more likely that one of the other possibilities suggested by Evans was responsible, or perhaps her conventional medication was making her worse and her decision to stop it was right. I tried thuja on many other psychiatric patients over the subsequent 20 years but not one of them responded.

The cases I've just described illustrate the difficulty of proving cause and effect in medicine, particularly when the origin of a patient's symptoms is obscure and the basis of the treatment you are using is questionable. Even when the treatment appears to work, this does not prove that the theory it relies on is correct. This is true of both psychotherapy and homeopathy.

Much or all of homeopathic theory may be mistaken, and the remedies themselves may have little objective efficacy or none at all, but patients often get better nevertheless. To say that this is due to the placebo effect is to beg the question, because we have only hazy notions about how placebos work anyway. For many patients, especially those whose symptoms really arise from their life situation, merely stating their problems verbally is sometimes enough to put them in a new light and to suggest the direction to look for a solution. In such cases the therapist is merely a sounding board; indeed, even a computer will do as a listener for some people. Many others do need a human individual to interact with, however.

So is the therapist no more than a sympathetic friend? No; this is where the theory comes in. It often doesn't matter much what a therapist's theoretical beliefs are (provided they are not actually dangerous, of course); their function in many cases is not to be "right" but to provide a framework to keep the discussion in focus.

The homeopath is not just chatting vaguely and asking questions at random, but is trying to use what the patient is saying as a guide to the right remedy. This gives the interview a frame of reference and prevents it from becoming totally shapeless. In this sense, homeopathy undoubtedly "works".

Most practising homeopaths, of course, would reject this analysis and would insist that the remedies they use have real effects and that the psychotherapeutic aspect of the consultation is secondary. And their belief in homeopathy is pretty well immune to any failures of research to support their position, because what they practise is, for them, more than a mere medical theory. It provides a frame of meaning that is likely to ensure the survival of homeopathy for a long time to come.

The future of homeopathy

Homeopathy has been with us for 200 years and has survived in spite of at times venomous attacks by orthodox doctors, so it certainly has staying power. At one time, at least in Britain, it was used almost exclusively by a small band of middle or upper class devotees, and few people outside this circle had heard of it. Today it is part of a wider and seemingly unstoppable wave of public enthusiasm for all kinds of unconventional medicine. Research is being carried out with the aim of justifying it, but the fact remains that, for many of its enthusiasts, the real point of it is precisely that it is *not* the same as conventional medicine. Much of the popularity of alternative medicine today, homeopathy included, is due to its philosophical difference from mainstream medicine. The fact that it is condemned as unscientific by some orthodox doctors is for many of its adherents a positive merit, not a criticism.

As we saw earlier, Richard Hughes in the nineteenth century tried to bring homeopathy and orthodox medicine together. If he had succeeded homeopathy would have lost much of its aura of mysticism and ultimately become just a branch of pharmacology. If that ever happens those homeopaths who are reacting against science (most non-medical homeopaths but some doctors too) will lose interest in it and look elsewhere for what they need.

Like other kinds of complementary–alternative medicine, often called CAM today, homeopathy is described as "holistic" by its defenders, in contrast to orthodox medicine, which is characterised as "reductionist". To some extent, I think, this is a bogus distinction. For one thing, nearly all CAM is practised privately, and it is a great deal easier to be "holistic", entering deeply into a patient's emotional problems, when you know there are not several other patients drumming their fingers outside your door. For another, good orthodox doctors do try to take account of the wider aspects of their patients' problems, in so far as patient numbers — and the "targets" imposed by a host of administrators — allow them to. Despite what they claim, CAM practitioners don't have a monopoly in respect of "holism".

What, then, is "holistic" medicine? It is more than advising patients about their diet, life choices, exercise and so on. It does include these things but there is something else. What really makes a medical system "holistic" is that it is based, implicitly and sometimes explicitly, on a "spiritual" view of human nature.

In CAM circles we often hear talk of a tripartite arrangement: body, mind, and spirit. CAM is supposed to treat all three. Orthodox medicine, in contrast, is founded on materialism. Individual doctors can of course have religious beliefs like anyone else but these are extra, so to speak; it is no part of a medical training to indoctrinate students with any kind of religious or spiritual view of human nature. Indeed, the usually unspoken implication of the modern scientific attitude is that human beings are (just?) complicated physiological mechanisms.

Medical training today is based on science. Even to get into medical school a young man or woman is expected to have gained good marks in science subjects, and science will continue to predominate throughout his or her medical training. The more able students may take a year off their main medical course to work for a BSc degree, and those who hope for a consultant appointment after qualifying will find themselves more or less obliged to do some research, whether or not they have any aptitude for it; one of the first things that appointment committees look at is usually the candidates' publication lists.

Some people deplore this emphasis on science but it is unavoidable. Modern doctors depend critically on science for both diagnosis and treatment. They must use the results of blood tests, bacteriology, x-rays, imaging, and a host of other investigations — the range is growing constantly — and they have to be able to understand these techniques at least to the extent of knowing how to select and interpret them sensibly. They must also have some knowledge of chemistry and pharmacology to help them choose the best treatment and watch out for unwanted effects. They are expected to maintain a critical scientific attitude to claims made on behalf of new treatments, to "keep up with the literature", and above all to practise "evidence-based medicine".

Critics of modern medicine usually object that it concentrates much too exclusively on identifying problems that have a convenient technological solution and ignores everything else. Doctors, it is alleged, are trained to have a garage mechanic's attitude to patients: find out what the knocking noise in the engine is due to and adjust or replace parts as necessary. But, the argument continues, human beings are not motor cars and in any case the whole idea of specifying problems is too narrow-minded.

It's difficult to deny the force of this argument. It takes only a little practical experience of medicine to reslise that many of the problems patients bring to their doctor are not susceptible to cure by technology — not even the technology of an idealised future. Many patients are old, or poor, or simply unhappy. They may come to the doctor with physical complaints — backache, headache, bowel problems — but the doctor knows in advance that medical tests are unlikely to turn up any physical cause for these symptoms. In many cases they are really expressions of an underlying unhappiness and are due to what is called somatisation. They are produced because it is more socially acceptable to complain of a physical symptom than of unhappiness, and also because simple unhappiness is not usually regarded as within the scope of medical treatment. And even when there is a physical cause for the symptoms they are often used as a pretext. We all know people who continue with their lives uncomplainingly in spite of considerable suffering or disability, while others, less severely affected, make their physical problems the focus of their attention and an excuse for receiving special consideration at work or at home.

Quite often the only solution a busy doctor can find for such patients is to prescribe an antidepressant, though in many cases a placebo might be as good or better. So-called tranquillisers are another possibility; it's said that at one time some high-pressure American companies provided bowls of tranquilliser capsules at meetings so that participants could take a few whenever they felt the stress was becoming too much for them.

For homeopaths, as for other CAM practitioners, this whole approach is wrong. Antidepressants, tranquillisers, and similar drugs treat the symptoms but not the root of the problem, they say, and this is because mainstream medicine doesn't look at the whole picture. It's interesting that the word used by many CAM practitioners to describe their work is "healing". The word means, literally, "making whole", and so it links with the claim that CAM is "holistic". Orthodox ("allopathic") doctors, it is said, are not healers in this sense, nor are hospitals healing shrines. In the past things were different.

The physician as healer

One of my favourite places in Greece is the site of the Sanctuary of Amphiaraeion. It lies in a fold of the mountains of Attica, about seven miles inland from the sea. Even today it is a magical spot. Set amid dense woods, with a little stream running through it, it is nearly always peaceful. Few tourists seem to know of its existence or to think it worth the trouble of a visit, so there are no guided tours to disturb the tranquillity; no postcards are on sale. Sitting there quietly alone you can persuade yourself that you are experiencing a little of the atmosphere of peace and healing that must have characterised the place in antiquity.

For the Sanctuary was a shrine to which patients came to be cured. It commemorates the elevation to divinity of Amphiarao, the great seer and warrior of Argos, who fought as one of the Seven against Thebes. When this expedition was defeated Amphiarao fled and was swallowed up, together with his chariot, near Thebes. As was the practice at other places of healing in ancient Greece, a patient wishing to consult the god would sacrifice a ram and lie down for the night, wrapped in its skin, in a special portico to await a divine communication in the form of a dream. After his cure he had to throw gold or silver coins into the sacred spring; sometimes grateful patients made votive offerings in the shape of the parts of their bodies that had been afflicted.

This method of seeking healing persisted for many centuries in Greece (in fact, resorting to shrines for healing and making votive offerings still continues in modern Greece, though now in a Christian context). Rather similar sites existed in other parts of the ancient world. They didn't always work, of course; one sceptical writer in late antiquity remarked that if all the people who had *not* been cured had presented votive offerings the shrines could not have contained them. Still, the fact that the practice survived for as long as it did shows that it must have worked for some, and the idea of using dreams for healing is by no means dead today, for it forms a cornerstone of Jungian psychotherapy.

It's easy to dismiss healing of this kind as relying on the "placebo effect", as if that made it unreal or insignificant. But healing can have an important place even when there is identifiable physical disease, while for the large number of patients whose symptoms lack an obvious physical cause healing may be all that is available.

It may well be that a great deal of healing is really self-healing by the patient and that the function of the therapist is mainly to enable this to occur. This is by no means to discount the role of the healer, quite the contrary, but we need to keep a sense of proportion and to reslise how little we really know. Before the modern era physicians had to be healers because hardly any of their treatments really worked. Today the opposite is the case; physicians have many treatments that work and the old-fashioned "good bedside manner" is thought of as little better than quackery. One might expect that psychiatrists would be the last medical healers, but they, too, have sold the pass. Modern psychiatry is increasingly becoming a branch of neurology, as an editorial in *The Lancet* pointed out a few years ago. It is CAM that is the real refuge of the healer today.

Modern orthodox doctors find themselves the uneasy heirs to two quite different traditions. On the one hand they are trained as scientists, they use scientific concepts, their tools and medicines are provided by science, and indeed their patients expect them to be scientists. On the other hand they are also expected to be healers, initiates of ancient mysteries who can provide answers to the deepest questions of life and death. No one human being can combine these roles fully and probably the second role cannot be adequately filled by anyone today. It seems often to be those doctors who were initially attracted to medicine because they saw themselves as healers who later take up homeopathy.

Homeopathy and the healer

For some of those who practise it homeopathy is more than just a method of using medicines. The announcement of a forthcoming lecture (the Richard Hughes Memorial Lecture, ironically) asks: "What can we learn from the homeopathic materia medica that we can use to live better, more fulfilled lives?" Try replacing "the homeopathic materia medica" with "antibiotics" to see the oddity of the question. But this is nothing compared with the claims that some homeopaths are prepared to make. Here, for example, is a prominent and influential non-medically-qualified homeopath, George Vithoulkas (my italics):

> It is absolutely certain, and every visionary man and woman is sensing it, that medicine today stands on the threshold of a deep and radical change and that soon it will embrace the new and unique possibilities that homeopathy is offering it ... It is my strong belief and my experience that homeopathy can effectively help ailing humanity in this endeavour and be an invaluable asset *for a speedy spiritual evolution of mankind.*

"Spiritual evolution of mankind"? The millenarian element that characterised nineteenth-century homeopathy in America is evidently still alive and kicking. Homeopathy is, for at least some of its more extreme advocates, more than a mere medical system — it almost seems to be taking on the role of a religion. Many homeopaths have what has been called the religious temperament and it colours their thinking about homeopathy.

Homeopathy is not a metaphysical system as such, of course, but it does have metaphysical elements and this is an important part of its appeal for some. It is, I believe, impossible to understand it fully unless we take this into account. When reading the great metaphysicians such as Spinoza or Kant we have to enter into their world for a time if we are to get anything useful out of the exercise; this requires the use of the sympathetic imagination (Keats's "negative capability"). I think we have to do the same with homeopathy if we are to understand why some people feel so passionately about it.

It is perfectly possible to practise homeopathy without taking account of any of this. Nineteenth-century British homeopathy under Hughes and Dudgeon did have this pragmatic character, and that is still the case today for many of those who use it. especially if they are medically qualified. But I don't think we can understand the full appeal of homeopathy for a lot of enthusiasts unless we reslise that it has this extra dimension. The fact that it does will almost certainly ensure that it continues to flourish no matter what objections are raised by scientifically-minded critics. Love it or loathe it, homeopathy is here to stay.

Appendix A

Books cited in the text

1. Allen TF. The Encyclopaedia of Pure Materia Medica, vols 1-10, 1874.

2. Buranelli, V. The Wizard from Vienna: Franz Anton Mesmer and the origins of hypnotism. London: Peter Owen, 1973.

3. Cook TM. Samuel Hahnemann: The Founder Of Homeopathic Medicine. Wellingborough, 1981.

4. Dudgeon RE. Lectures On The Theory And Practice Of Homeopathy. Manchester, 1854.

5. Evans, D. Placebo: The Belief Effect. London: HarperCollins, 2003.

6. Forrest, D. The Evolution of Hypnotism. Forfar: Black Ace Books, 1999.

7. Gauld, A. History of Hypnotism. Cambridge: Cambridge University Press, 1995.

8. Haehl R. Samuel Hahnemann, His Life And Work. London, 1922. (Indian edition)

9. Hahnemann S. The Chronic Diseases; Their Specific Nature and Homeopathic Treatment. Tr. Hempel. New York, 1845.

10. Hahnemann S. Lesser Writings. Tr Dudgeon. (Indian edition)

11. Hahnemann S. Materia Medica Pura, vols. 1-4. (Indian edition)

12. Hahnemann S. The Organon Of Medicine. Tr Dudgeon.1970: Indian edition.

13. Hughes R. The Principles and Practice of Homeopathy. (Indian edition)

14. Hughes R. Cyclopaedia of Drug Pathogens, vols. 1-4. London, 1886-91.

15. Kaminer, W. Sleeping with Extraterrestrials: The rise of irrationalism and perils of piety. New York: Vintage Books, 1999.

16. Kent JT. Lectures On Homeopathic Materia Medica. Indian edition, 1966.

17. Kent JT. Lectures On Homeopathic Philosophy. Chicago, 1919.

18. Kent JT. Repertory Of The Homeopathic Materia Medica. Indian edition, 1969.

19. Schiff M. The Memory of Water: homeopathy and the battle of ideas in the new science. Thorsons, 1995.

20. Storr, A. Feet of Clay: The power and charisma of gurus. New York: Free Press, 1996.

21. Tyler M. Homeopathic Drug Pictures. Saffron Walden,1952.

22. Vithoulkas, G. The Science of Homeopathy. New York, Grove Press, 1980.

23. Wood M. The Magical Staff: The Vitalist Tradition In Western Medicine. Berkeley, 1996.

Appendix B

Reviews and meta-analyses cited

1. Cucherat M. et al. Evidence for clinical efficacy of homeopathy. A meta-analysis of clinical trials. European Journal of Clinical Pharmacology 2000;56(1):27-33.

2. Ernst E. A systematic review of systematic reviews of homeopathy. British Journal of Clinical Pharmacology 2002;54(6):577-582.

3. Feder, G, Katz, T. Editorial: Randomised controlled trials for homeopathy: Who wants to know the results? BMJ 2002;324;498-499.

4. Jonas WB et al. A critical overview of homeopathy. Annals of Internal Medicine 2003;138(5):393-399.

5. Kleinen J, Knipschild P, ter Riet G. Clinical trials of homeopathy. Br Med J 1991;302:302-23.

6. Linde K, Clausius N, Ramirez G et al. Are the clinical effects of homeopathy placebo effects? A meta-analysis of randomised controlled trials. Lancet 1997;350:834-843.

7. Shang A, Huwiler-Müntener K, Nartey L et al. Are the clinical effects of homeopathy placebo effects? Comparative study of placebo-controlled trials of homeopathy and allopathy. Lancet 2005;366:726-732.

Appendix C

Modern provings cited

1. Clover AM et al. Report on a proving of pulsatilla 3x. Br Homeopath J 1980;69(3):134-139.

2. Walach H. Letter: Br Hom J 1996;85:123-25.

3. Walach H, Koster H Hennig T, Haag G. The effects of homeopathic belladonna 30CH in healthy volunteers — a randomized, double-blind experiment. J Psychosom Res 2001;50(3):155-60.

Appendix D

Clinical trials cited

1. See the **Faculty of Homeopathy:** Randomised controlled trials in homeopathy (www.facultyofhomeopathy.org/research/randomised-controlled-trials-in-homeopathy/). Note that this page no longer gives details of the studies alluded to in the text but now simply lists a large number of published papers. Nevertheless I have continued to include a discussion of the studies previously cited by the Faculty because they give a useful indication of the kind of problems one encounters in trying to assess research in homeopathy.

2. de Lange de Klerk ESM, Blommers J, Kuik DJ, et al. Effect of homeopathic medicines on daily burden of symptoms in children with recurrent upper respiratory tract infections. Br Med J 1994; 309: 1329-32.

3. Ferley JP, Zmirou D, Adhemar D, Balducci F. A controlled evaluation of a homeopathic preparation in the treatment of influenza-like syndromes. Br J Clin Pharmacol 1989; 27: 329-35.

4. Harrison H, Fixsen A, Vickers A. A randomized comparison of homoeopathic and standard care for the treatment of glue ear in children. Complement Ther Med 1999; 7: 132-5.

5. Lewith GT and others. Use of ultramolecular potencies of allergen to treat asthmatic people allergic to house dust

155

mite: double blind randomised controlled clinical trial. BMJ 2002;324:520-3.

6. Reilly D, Taylor MA, Beattie NGM, et al. Is evidence for home-opathy reproducible? Lancet 1994; 344: 1601-6.

7. Sam C, Bonnet MS, van Haselen RA. The efficacy and safety of a homeopathic gel in the treatment of acute low back pain: a multi-centre, randomised, double-blind comparative clinical trial. Br Homeopath J 2001; 90: 21-8.

8. Walach H et al. Classical homeopathic treatment of chronic headaches. A double-blind, randomized placebo-controlled trial. Cephalalgia 1997;17:119-26.

9. Walach H et al. The long-term homeopathic treatment of chronic headaches: one year follow-up and single case time series analysis. Br Hom J 2001;90:63-72.

10. White A, Slade P, Hunt C et al. Individualized homeopathy as an adjunct in the treatment of of childhood asthma: a ran-domised placebo controlled trial. Thorax 2003;58:317-21.

Appendix E

Potency

1. Belon E, Cumps J, Ennis M et al. Histamine dilutions modulate basophil activation. Inflamm res 2004;53:181-88.

2. Homeopathy 2007;96(3). Special Issue: The Memory of Water. [This has a range of articles discussing ideas about how the potency effect might work.]

3. Winston, J. A brief history of potentising machines. Br Hom J 1999;78:59-68.

Appendix F

Hyperventilation studies

1. Bass C. Hyperventilation syndrome: a chimera?. Journal of Psychosomatic Research 1997;42(5):421-6.

2. Hornsveld HK, Garssen B et al. Double-blind placebo-controlled study of the hyperventilation provocation test and the validity of the hyperventilation syndrome. Lancet 1996;348(9021):154-8.

3. Morgan WP. Hyperventilation syndrome: a review. American Industrial Hygiene Association Journal 1983;44(9):685-9.

4. Troosters T, Verstraete A et al. Physical performance of patients with numerous psychosomatic complaints suggestive of hyperventilation. European Respiratory Journal. 14(6):1314-9, 1999.

Appendix G

About placebos

1. Beecher HK. The powerful placebo. Journal of the American Medical Association 1955;159:1602-6.

2. Campbell A. Cartesian dualism and the concept of medical placebos. Journal of Consciousness Studies 1994;1:230-33.

3. Evans D. Placebo: The Belief Effect. London: HarperCollins, 2003.

4. Jones AKP, Kulkarni B, Derbyshire SWG. Functional Imaging of Pain Perception. Current Rheumatology Reports 2002;4(40;329-333.

5. Nitzan U, Lichtenherg P. Questionnaire survey on use of placebo. BMJ 2004;329:944-946.

6. Park LC., Covi L. Nonblind placebo trial. Archives of General Psychiatry 1965;12:336-345

7. Porro CA. Functional Imaging and Pain: Behaviour, Perception and Modulation. The Neuroscientist 2003;9(5):354-69.

Index

www.ingramcontent.com/pod-product-compliance
Lightning Source LLC
Chambersburg PA
CBHW032015170526
45157CB00002B/708